WONDER AND DELIGHT

'Instead of an intense emotional conviction that his view of Nature is the right one at last, the scientist's chief feeling is one of enjoying finding out and enjoying gaining a wider understanding of Nature. The growth of knowledge is his main concern, not its storage. He shares with our earliest ancestors a sense of curiosity and feelings of wonder and delight; and he extends these into a great sense of intellectual progress. You can share that with him.'

Eric Rogers

Physics for the Inquiring Mind p 758
Princeton University Press 1960

Eric M Rogers
Photograph reproduced by permission of AIP Emilio Segrè
Visual Archives (Physics Today Collection)

WONDER AND DELIGHT
Essays in Science Education in honour of the life and work of Eric Rogers 1902–1990

Edited by

Brenda Jennison and Jon Ogborn

With a Preface by

John L Lewis OBE

Institute of Physics Publishing
Bristol and Philadelphia

British Library Cataloguing-in-Publication Data

A catalogue record for this book is available from the British Library.

ISBN 0-7503-0315-8

Library of Congress Cataloging-in-Publication Data are available

Published by Institute of Physics Publishing, wholly owned by The Institute of Physics, London

Institute of Physics Publishing, Techno House, Redcliffe Way, Bristol BS1 6NX, UK

US Editorial Office: Institute of Physics Publishing, The Public Ledger Building, Suite 1035, Independence Square, Philadelphia, PA 19106, USA

Printed and bound in Great Britain by Bookcraft Ltd, Bath

CONTENTS

CONTRIBUTORS

Susana de Souza Barros: Federal University of Rio de Janeiro, Rio de Janeiro, Brazil

Paul Black: King's College, University of London, UK

Joan Bliss: King's College, University of London, UK

Bryan Chapman: Centre for Studies in Science and Mathematics Education, University of Leeds, UK

Maurice Ebison: formerly Institute of Physics, London, UK

Marcos da Fonseca Elia: Federal University of Rio de Janeiro, Rio de Janeiro, Brazil

Anthony French: Massachusetts Institute of Technology, Cambridge, Mass., USA

Keith Fuller: formerly Bedales School, Petersfield, UK

Rachel Gevertz: Ministry of Science and Technology, Brazil

Tim Hickson: The King's School, Worcester, UK

Jim Jardine: formerly Moray House, Edinburgh, Scotland

Brenda Jennison: Department of Education, University of Cambridge, UK

Nahum Joel: formerly University of Chile and UNESCO, Paris, France

Goronwy Jones: formerly Department of Education, University College Cardiff, Wales

John Lewis: formerly Malvern College, Malvern, UK

Ana Tereza Filipecki Martins: SENAI, CETIQT, Rio de Janeiro, Brazil

George Marx: Department of Atomic Physics, Eötvös University, Budapest, Hungary

Jon Ogborn: Institute of Education, University of London, UK

Tae Ryu: Sophia University, Tokyo, Japan

Joan Solomon: University of Oxford, Oxford, UK

Charles Taylor: formerly Department of Physics, University College Cardiff, Wales

Poul Thomsen: Royal Danish School of Educational Studies, Copenhagen, Denmark

Eszter Tóth: Lauder Yavne School, Budapest, Hungary

together with selections from Eric Rogers' own writings and lectures.

PREFACE

Ruthless, yet at times very gentle and caring—egotistical, demanding, yet often very entertaining and sometimes sentimental—always stimulating, sometimes infuriating and on occasions totally impossible to work with—in other words, Eric Rogers was very human and a mass of contradictions. Sometimes he was childlike in his simplicity, which probably explains why he could have such rapport with young people. Sometimes he was Machiavellian in his cunning when he wanted his own way. Of course he always enjoyed an audience as there was a lot of the actor within him. In many ways he was a very lovable person who inspired devoted support, and there is no doubt that he made a major contribution to physics education, not only in the UK, but throughout the world.

My first sight of Eric Rogers was on a screen at a PSSC Summer Institute in the University of Colorado. The American teachers were as excited by the film *Coulomb's Law* as I was. This lively man leaping from apparatus to apparatus with all the skill of a professional actor, was capable of riveting totally an audience's attention: it was an object lesson for any teacher.

I first got to know him well when together we attended an IUPAP conference in Rio de Janeiro in July 1962. This was a time of ferment in science education, which was rapidly becoming a worldwide concern. Pioneers such as Zacharias, Holton and Feynman were there and the widespread respect and affection for Eric Rogers was obvious. Such was his inquiring mind that within hours of arrival he had unearthed a basement exit from our hotel so that he could make a 6 am exit to Copacabana beach each morning for a swim in the rolling breakers, despite the pollution.

His cunning was revealed on the journey home. His wife Janet was a sculptress and Eric had searched Rio for a particular block of hardwood. He had no intention of paying extra to bring back this 40 lb block. It was therefore enclosed within his raincoat. He explained his system. The word 'Professor' should be clearly stated in his passport, the block was enclosed in his raincoat which he let trail on the ground and there should, if possible, be a German businessman behind him in the queue. He would become very vague when checking in, producing all the wrong papers, together with a couple of innocuous parcels. Of course the cunning old actor was successful: in desperation they bundled us through with the block unnoticed.

He was very difficult to work with—as was found by his team, by the Foundation, by the apparatus manufacturers and by the publisher—but it was a rewarding experience for those of us who had the privilege. He had his influence on the American Physical Science Study Committee; *Physics for the Inquiring Mind* remains an invaluable resource, rich in wisdom. He made a profound contribution to the Nuffield projects, which were valued and enjoyed by enormous numbers of young people, and have had even wider influence on science teaching in all parts of the world. This book will testify to a very great man. We all owe much to his genius.

John L Lewis

EDITORS' INTRODUCTION

Eric Rogers was a great physics teacher, with an international reputation for the passion, quirkiness and profundity of his thinking. The idea for a book in his memory was born out of discussions in the Education Group of the Institute of Physics, and between the editors and Paul Black, Bryan Chapman, Brian Davies, Jim Jardine, Goronwy Jones and John Lewis.

It was decided that the best way we could honour Eric Rogers' memory would be to collect together a book of writings about science education, on subjects about which he cared deeply, written by an international group of distinguished authors, each of whom had something of importance to say which would be of wide and lasting interest. We invited contributors to write anything which they thought might have interested Eric Rogers.

In Part I, eleven substantial essays deal with matters of current and lasting concern in science education. They are all matters which Eric Rogers had close to his heart and to which he made important contributions. George Marx from Hungary gives the book an immediate international perspective, taking a broad historical canvas and vividly evoking the issues which confront the need for science education to provide a path towards the future, particularly for developing countries. Jon Ogborn writes about what we need to do if we want to have a science curriculum which is truly for all the people. Joan Solomon describes how an understanding of the nature of science can be helped by having pupils discuss ideas taken from the history of science. Joan Bliss surveys and criticizes several decades of thinking about the problems of learning science, illuminating relatively unknown aspects of the work of Piaget, and suggesting how we can now best develop our understanding of how children understand the world. Paul Black points the way towards a humane and helpful assessment system, drawing on a wide range of evidence and providing arguments for those who confront authorities who do not see both how practical and how right is the demand for constructive humanity in assessment. Tae Ryu is an honorary Nuffield physics teacher who provides a Japanese viewpoint, describing Eric Rogers' influence on teaching and examining in Japan but also pointing to the difficulties which can be caused by an inflexible examination system. Anthony French gives an American perspective, describing an imaginative experimental physics course at MIT based on 'take-home' experiments, to replace and improve on conventional laboratory work, and showing that physics can be taught through simple 'string and sealing wax' experiments. Charles Taylor, a master of the art of demonstrating to audiences aged from 5 to 95, offers advice and wisdom about demonstration. Jon Ogborn adds a chapter reflecting Eric Rogers' concern with models in science, but now seeing what kind of modelling clay the computer might provide. Maurice Ebison returns to a theme which was always central to Eric Rogers' concerns, the understanding of how science grows, which can be provided by a discussion of ancient astronomy and theories of the universe. Eszter Tóth then closes the Hungarian bracket around Part I with an account of how she and her pupils became involved in a real and urgent problem of radioactivity and in the process learned much more than some scientific information.

Parts II and III both concern Eric Rogers more directly.

Part II gives examples of Eric Rogers' influence on physics education, through his own words and through examples given by others. It begins with Eric Rogers' striking Oersted Medal address, in facsimile, containing one of the most powerful statements of his thinking about tests and examinations. His address to the Edinburgh ICPE Conference in 1975, transcribed and edited for this volume, is even more personal, providing an excellent example of his authentic voice. John Lewis, who himself played an essential role in the Nuffield O-level physics project, follows with an account of the development of the project and of Eric Rogers' crucial part in it. Tim Hickson records the impact some of the ideas from the project made on him as a classroom teacher. Jim Jardine records, with some historic illustrations, central aspects of the development of apparatus for pupils to use in investigations and to see in demonstrations. Nahum Joel returns us to the international scene, describing the impact in South America of Eric Rogers' 'question shredding' approach to getting people to reflect on and re-think their aims in science education. Bryan Chapman imagines the scorn and ridicule Eric Rogers might have poured on the British national curriculum, were he to return in the year 2000. To conclude, Brenda Jennison looks back at the tremendous value a young trials teacher got out of her interaction with Eric Rogers and what he stood for, and she ends by regretting recent failures to learn the constructive lessons of those days.

Part III is even closer to the man, with personal accounts of Eric Rogers, plain spoken as he would have wished and expected. They come from different countries; from Anthony French in the USA, from Poul Thomsen in Denmark, from Goronwy Jones in Wales and from Susana de Souza Barros, Marcos da Fonseca Elia, Rachel Gevertz and Ana Tereza Filipecki Martins in Brazil. The effect he had on those who knew him is vividly evoked. Keith Fuller provides a brief biography, which gives particular insight into the influence on Eric Rogers of Bedales School and its first headmaster. There follow three citations, for the Oersted Medal in 1969, the ICPE Medal in 1980 and the Bragg Medal in 1985.

Einstein's friendship with Eric Rogers is celebrated with a charming tale of Einstein's birthday present from him. Finally Eric Rogers, as he always did in life, gets the last word. We have chosen *The demon theory of friction* as an example of his ability to combine wicked wit with the making of a serious point.

Eric Rogers' voice also speaks throughout the book. The various contributions are introduced with short quotations from his writings. They are powerfully evocative for those who knew him, and to those who did not will convey better than anything else in this book the essential nature of the man. Re-reading *Physics for the Inquiring Mind* all through for the first time for many years, we were struck by how little it has dated. Already in the 1950s he was drawing attention to the problems of world energy resources. His explanations are as fresh and full of insight today as the day they were written. If you have not read it recently, please do so.

It has been a privilege to edit this book. We thank the many distinguished people who responded to our requests for contributions and who then accepted our sometimes heavy editorial hand with good grace. Thanks also go to Debbie Radford who typed up all the manuscripts we could not get in electronic form.

This book is, in many ways, a call to arms. Science education has become politicized. Science education has to respond, and from within. Let us re-assert the vigour of science as a part of human culture. We have nothing to lose but our brains.

Brenda Jennison
Jon Ogborn

PART 1

ESSAYS IN SCIENCE EDUCATION

A fable of conferences on physics teaching

The first conference—probably promoted by GIREP and UNESCO— was held in the year 1700. At that conference, the discussion of school teaching of physics reached excellent agreement: that the teachers know very well that when we apply a force an object moves with constant velocity, and that double the force maintains double the speed. They have a good experimental demonstration: a tall jar of oil in which they release a tiny glass ball which descends with constant speed; then a tiny iron ball which descends with greater constant speed. Everything is in good order. The conference voted that such teaching should continue and that nothing should be changed—except the teachers' salaries— and, as regards Newton's Laws, the work of that man Newton should be reserved for advanced graduate seminars in universities.

A century later, around 1840, GIREP again organized a conference on physics teaching. Newton's Laws are being taught very well and are used for the purpose Newton expected: magnificent astronomy. Caloric is also well known to all the teachers as something that is completely conserved, explaining many experiments: mixing hot and cold water, throwing hot lead shot into water, and sliding down a rope and squeezing caloric out of it. The conference decided that the Conservation of Caloric should continue to be taught. As for the idea of heat as energy, the strange suggestion from the young man Joule in England and from Mayer in Germany, that should be reserved for advanced lectures in universities—with a severe warning that it is probably nonsense.

In 1910 there was yet another physics teaching conference. Newtonian mechanics is well established, and in good use except that the French mathematicians have made it look more difficult. General conservation of energy is also accepted. But the strange idea of Planck (which he himself did not wholly like) should be reserved for special post-doctoral seminars at universities. So also should the unfortunate attack on geometry made by Einstein, which makes people feel so uncomfortable.

The first part of my story is imaginary, but the ending is real. The question is: are we going to be as far behind—by half a century or a century—in the year 2000? Well, why not? In 1700 there was good teaching of pulleys, of hydrostatics—mostly pieces of physics which the Greeks discovered. And even today we teach a lot of useful physics, so why worry about the delay?

(continued)

Because of the rate of communication! We can travel round the world in a matter of days instead of months; we can send messages around the world in seconds; we can even go to the Moon. Scientific work is being done not by a selected few with a special interest, like the noble Robert Boyle or the 'Nobel' Niels Bohr. We are not just educating future professors of physics—they are fool-proof anyway and will develop almost independently of our teaching. We are teaching physics to future technicians, to people in other sciences, and—above all—to the general educated public. If we are far behind the times in our teaching, we shall be doing a damage to public understanding. So I make a strong plea that we try to accelerate.

Eric Rogers *Proceedings of GIREP Conference, Copenhagen, 1969*

CHAPTER 1

SHORTCUT TO THE FUTURE

George Marx

Our grandchildren are our immortality
Janet Tran Rogers

1.1. ACCELERATING HISTORY

In antiquity, changes happened very slowly as compared to the rate of change of generations. *Time seemed to stand still*. In this era man created timeless models for the world he lived in. Dynasties ruled over 'eternal' empires by the grace of God, whose moral commandments were carved in marble. The ultimate goal of rulers and believers was immortality. *The model of man was the statue*. Timeless arts like architecture and sculpture expressed the overall desire to realize *eternity*. The main contribution of antiquity to science was the timeless world model called *geometry*. These views survive even today in some school books, where they emphasize the static equilibrium of forces, rigid bodies, fixed stars, indestructible atoms and unchangeable species. Motion was an outlaw, change was considered harmful, and time was an ugly concept to be eliminated from the world order. For a stable society this was a very efficient cultural model, born in the *Mediterranean*, and serving well through several millennia up until the Renaissance. In this situation, the reproduction of society relied upon *family education*, in which skills were transmitted by imitation of the parents. For the young, the *father* was the ultimate authority.

Five hundred years ago, Columbus discovered a New World. Geographical expansion, world trade and the industrial revolution redesigned the shape of the world. According to the baroque taste, *motion is beautiful*. Arts such as *music* and *theatre* in time expressed the soul of the New Age: Monteverdi and Bach, Shakespeare and Molière were the new heroes of beauty (preparing the taste for movies, radio and television). Progress had become fashionable: Marx wrote about class struggles and *revolutions*; Toynbee emphasized the call of *new frontiers*. These more efficient models shifted the centre of gravity of history from the Mediterranean basin— from Italy, from Islam—to the coasts of the *Atlantic*, where Spain and Portugal ruled over the oceans, followed by England and the Netherlands. In Rome, Galileo was pressed by the Inquisition to withdraw his dangerous teachings, but after the process still insisted, *'Eppur si muove!'* (In 1992 the Catholic church acknowledged that the Inquisition had been wrong and that Galileo was right.) Even a *reformation* of the church occurred, starting in the industrializing countries. Newton and Darwin were buried in Westminster Abbey in tombs surpassing in grandeur those of the English kings. To science, Galileo and Newton brought dynamics and Darwin brought a dynamics of living species. *The steam engine served as the new model for man*, transforming fuel to motion. By today, this historic point of view has penetrated even the exact sciences: continents drifted, the climate changed, the planetary system was formed, and even the whole universe started with a Big Bang. People moved from farms to factories. Imitation of the father was no longer enough for social survival. Society invented the *school*, where an updated knowledge was to be transferred to the next generation. Curricula were revised and school books were re-written from time to time. Instead of the father, the *teacher* became the ultimate authority for fresh generations. By means of compulsory schooling, the rate of change was accelerated to become equal to the rate of change of generations. Older people started complaining: *In my time things were done very differently....*

In the second half of the twentieth century, modern science and its offspring—high technology—accelerated further the pace of history. Internal combustion (cars, aeroplanes), telecommunications (radio, television, telephone, fax, electronic mail), and computers and contraceptives have produced, within years, more lasting changes in the political–ethical–social web than any ideology, philosophy or religion could achieve in the previous centuries. According to the Hungarian–American John von Neumann *the computer is the appropriate model of man*, because it can be re-programmed, and because it possesses the capability of rapid handling of information. Reaching the Moon required not so much gigantic rockets as miniaturized electronics. World War 3 (the Cold War) was fought and won not by armies and bombs but by telecommunications and high technology. *Who are the winners of the Cold War?* asked Sergei Kapitza, president of the Russian Physical Society. He immediately answered his own question: *Japan and Germany*. If post-war national borders are to be shifted at all, it may well be at the Kuril Islands, and any such shift will not be forced by military might but by a superior economy. Competition between nations is conducted today through economies and education. When Russian citizens were asked what kind of help they expected from the USA (August 1991), the distribution of their answers was:

technological assistance　53%
financial aid　16%
food aid　14%
political assistance　3%
nothing　7% .

During human history, different models contributed to our cultural heritage:

China invented paper and printing, the compass, the rocket, and their astronomers were the first to sight supernovae. But the ultimate authority of the emperor's opinion stopped progress in the Middle Ages.

Europe discovered axioms of geometry, the concept of force, the regularity of planetary motion. But then they decided in favour of reading the classical authors—Aristotle and Ptolemy—and introduced the Holy Inquisition to fight the heretical idea of change.

India discovered zero, infinity, infinitesimal and irrational numbers. But the caste separation of craftsmen from thinkers (Brahmins) blocked technological progress.

Islam gave us the names of Algebra and of Aldebaran, and they invented optical lenses. But they stopped research 500 years ago.

Israel gave the world (including Christians and Muslims) a rich moral heritage. Scientists of Jewish origins excel and collect a large fraction of Nobel prizes. But the Israeli leaders are worried: why are Jewish scientists more successful in the Diaspora than in the Promised Land?

Japan discovered the meson and the tunnel diode, and gave people video-cameras. But she trains her students to pass closed-ended examinations, and has now got a creativity complex.

Russia gave us the periodic table, the Big Bang, and the concept of quasi-particles. But for ideological reasons she suppressed modern physics, molecular genetics and computer science, and so lost the high technology race in war and peace.

The USA presented mankind with automobiles, and then (50 years ago) nuclear power. She landed on the Moon. But the driving force of the American economy is marketing and advertising (the whole world knows the macho superiority of Marlboro and the youthful modernity of Pepsi). Although the majority of physics Nobel laureates work in the USA, they were educated in the schools of Europe and Asia. American society is adept at *using* talents. But American high schools and colleges are not able to *supply* the demand in scientific and engineering manpower: 50% of the PhDs for science in the USA go to foreigners, and this fraction may increase to 70% by the year 2000. In earlier decades the deficit was covered by brain imports from Western Europe; now it comes from Eastern Europe and from the Far East. This brain drain gained the nickname, *Reverse Marshall Aid*, and is driven by the difference in level of schools in the Old and New Worlds. As the semi-official report *A Nation at Risk* (1983) put it: *If an unfriendly power had attempted to impose on America the mediocre educational performance that exists today, we might well have viewed it as an act of war. As it stands, we have allowed this to happen to ourselves.* In his proclamation America 2000, President Bush invited the nation to get American schools to the top in science and mathematics by the year 2000. The Department of Education is investing two billion dollars, mainly in re-training science teachers. This could become an investment comparable to the Manhattan Project of President Roosevelt in the 1940s or to the Apollo Project of President Kennedy in the 1960s. The American Association for the Advancement of Science calls its Education Plan for a Changing Future *Project 2061*, reminding us that the kids of the present will see the next return of Halley's comet.

The fact that the rate of scientific–technological change accelerated until it was faster than the rate of change of generations has led to social tensions. In the past decades the whole social engine of the Atlantic culture has begun to show signs of trouble. *What Johnny has learned in school is not enough for Uncle John to prosper in his later years*, is the proverbial reason given for unemployment, caused by the restructuring of production from heavy industry towards electronics. Trade Unions behave like modern-day Luddites, trying to slow down progress, and advocating traditional coal mining and manufacturing instead of nuclear power and automation. But, as the pioneer of the computer era, John von Neumann, said, *There is no medicament against progress.*

The generation gap in the industrialized countries of the west is becoming wider and wider. Alienation, intolerance of and by youth, runaway kids, drugs, vandalism, tattooing and terrorism are just symptoms occurring amongst school-going children of well-to-do families. Previously, the school-leaving diploma was a licence not to have to learn any more. Now, teaching 'final truths'—appropriately called *dogmas*—may result in a loss of competitiveness of a nation. For teenagers, it is no longer their elders who are their idols, but instead the young *rock star*, who senses the trends and expresses the spirit of the Brave New World.

The traditional goal of school was the reproduction of society from generation to generation. Future-conscious nations have now to look for new social models. The Soviet Union spent 20% of its budget on its army. Korea spends more than 20% of the national budget on its schools. In Taiwan, the Constitution itself requires 15% of GNP, 25% of the local budget, and 35% of community expenditure, to be spent on education. The Ministries of International Trade and Industry in Japan and Korea support future-conscious developments; they consider quality of goods to be their main goal, and an export permit is only issued for products of the highest quality. Economists express the opinion that the *Pacific basin* may take over the role of the Mediterranean for the coming century.

1.2.   A BIOLOGICAL TALE

Let us consider a biological tale, to go with Eric Rogers' fable of accelerating change in physics teaching which is printed opposite the title of this chapter. What best lets change accelerate?

Suppose that x is any measurable property of a population (e.g. body mass of pigs, tail length of foxes, IQ of humans). Under natural conditions its distribution will have a Gaussian shape, with the maximum of the distribution at the value x_0 which gives the best fit to the environment, that is, the maximum rate of reproduction. Natural selection concentrates the distribution nearer to this value, reducing variation around it. By contrast, mutations widen the distribution. In dynamical equilibrium the effects of selection are just balanced by those of mutation. If the distribution in the nth generation is described by the normalized function $g_n(x)$, the average of x and its mean square deviation from the average are defined by:

$$x_n = \int x g_n(x)\, dx \qquad (\Delta x_n)^2 = \int (x - x_n)^2 g_n(x)\, dx.$$

In agriculture, human intervention tries to compress the distribution to the fastest breeding region by hybridization and other genetic tricks, that is, to compress $g(x)$ to the value $x = x_0$. Now let us imagine that an external change favours values of x larger than the old preference for x_0. Selection now shifts the distribution towards higher values of x:

$$s(x) = 1 + c(x - x_n).$$

c is a positive constant of selection so that $s(x)$ is positive if $x > x_n$ and negative if $x < x_n$. The distribution function in the next generation will be:

$$g_{n+1}(x) = s(x) g_n(x).$$

Let us calculate the average value of x in the new generation.

$$x_{n+1} = \int x g_{n+1}(x)\, dx = \int x[1 + c(x - x_n)]\, g_n(x)\, dx$$

$$= x_n + c \int (x - x_n)^2 g_n(x)\, dx + c x_n \int (x - x_n) g_n(x)\, dx$$

that is,

$$x_{n+1} = x_n + c(\Delta x_n)^2.$$

Under a stable climate, a small variability Δx has an advantage, with the whole population made of individuals close to the fittest variant with $x = x_0$. Under a variable climate, however, a wide spectrum is better, because one of the different individuals may have the appropriate answer to the new challenges. A larger Δx results in a faster increase of x, that is, in faster evolution.

1.3. SCIENCE EDUCATION IN A MOVING WORLD

1. PROFESSORS are asked to select the specific elements of scientific knowledge they consider to be necessary and long-lasting enough for citizens to need to learn.

2. SCHOOL TEXTBOOK AUTHORS chew on and describe these bits of information. Their method is reproduction; their main sources are previous textbooks.

3. TEACHERS transfer the selected factual knowledge to pupils by lecturing (information travels a one-way path).

4. PUPILS accept and memorize this information passively; in the examination they reproduce the words of the teacher and of the textbook. Then they quickly forget them, so as to have room for new transient data in the memory cells of the brain. The diploma they receive is a licence to stop learning.

5. SCIENCE, however, produces accelerating technological progress. Industrial and scientific revolutions come and go, influencing politics and the economy. School textbook authors try hard to keep pace by adding new pages to their books, but never mind—there is no time left to teach these extra pages. Thus a gap is created between active science and school science. Young people find school physics obsolete and irrelevant, and lose interest in scientific studies.

6. PROFESSORS start complaining about the decline of interest in science and the weak performance in university entrance examinations. They decide: GOTO 1.

These were my introductory words at a ICPE/IUPAP conference in Trieste in 1980, illustrated by the transparency shown in figure 1.1, and they attracted from Eric Rogers the highest praise of my life. He never stopped emphasizing the huge responsibility of science teaching. For a century, schools were the engine of social progress. Now they may become its brake.

But which of us can answer the following current questions:

Is *big* or *small* beautiful?
Is *extensive* or *intensive* development preferable?
Is *nostalgia* or the *avant garde* to be favoured?
Does *Mozart* or *The Beatles* best express the human soul?
Is *Flower Power* or *High Tech* for our benefit?
Is *city* or *country* life good for you?
Is *Back to Basics* or *Towards New Frontiers* the more promising slogan?
What fuel will drive cars—*sunshine* or *hydrogen*?
What will electricity come from—*coal* or *the nucleus*?

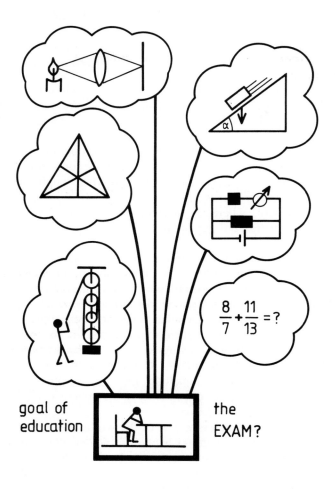

Figure 1.1. *Physics education for what?*

How will wars be fought—by *bombs* or by *trade*?

What will grow in Europe in the climate of the twenty first century—*wheat* or *dates*?

We *feel* that these alternatives contradict one another, but in some ways both are right, as are the *wave* and *corpuscular* models for the electron. Neither teachers, nor indeed adults generally, can take the responsibility of answering all these questions on behalf of the citizens of the future. Educators may discuss what to teach:

in mathematics: addition of fractions or computer languages?

in physics: sliding on inclined planes or the ozone hole?

in chemistry: manufacture of sulphuric acid or acid rain?
in biology: the lion or the AIDS virus?
in geography: Turkey or Korea?
in natural history: national parks or plate tectonics?
... and so on.

Our school heritage may prefer the first choices, while our newspaper-reading public may prefer the second. But both answers may turn out to be wrong, because the knowledge needed to survive in the twenty first century has probably not yet been discovered. What then is the delicate balance between *reproduction* and *invention*? Which do you prefer:

- the *normative* school, where the teacher tells the only true knowledge, where the teacher asks closed problems, and where pupils are rewarded for giving the expected answer, or
- the *creative* school, where the teacher offers alternative models, puts open-ended problems, and pupils are tolerated when they ask unexpected questions?

We have to educate our children to respect physical, economic and social realities *(observe!)*; to orient themselves in unknown situations *(understand!)*; to anticipate events which have never before occurred *(predict!)*, and, if the anticipation works, to act *(invent!)*. But this is just the way scientists work. Genuine scientists are happiest when their observations *disagree* with theoretical expectations: if so, this is the time to search for new models. This is why scientists are by profession sensitive to contradictions, changes and trends. This is why they tend to belong to the political opposition. Thus I conclude: the best education can do in this moving world is not to argue over items of the school curriculum but to offer *scientific literacy for all future citizens*.

Mathematics offers a timeless model of the world. *History* is in the past tense, good at explaining why things happened. *Geography* is in the present tense, good at describing present countries, borders and capitals. *Physics* is in the future tense, its goal being to anticipate events. It is this which makes physics relevant to the newly arriving citizens of our changing world. This is the point from which *technology* can start. As the Hungarian-born British Nobel laureate in physics, Dennis Gabor, put it: *The future cannot be predicted; it has to be invented*. And as Henry Brooks Adams said: *A teacher affects eternity; he can never tell where his influence stops*.

1.4. SEARCH FOR A NEW MODEL OF POST-INDUSTRIAL SOCIETY

At the turning points of history, some nations jump ahead while others lag behind. Ivan Berend, Hungarian vice-president of the International Society of Historians, reminds us: *The twenty first century has already*

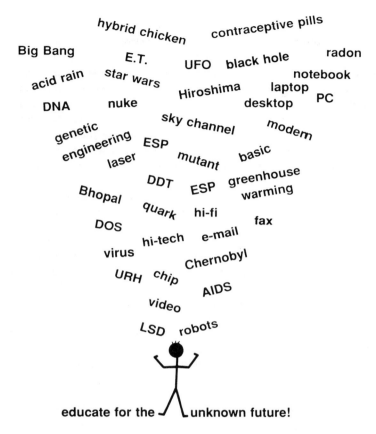

Figure 1.2. *The post-industrial world.*

begun in 1989. We are at a new turning point. The race has started. Some characteristics of the post-industrial era can already be discerned (see figure 1.2):

(1) As in avant garde novels, we shall experience *parallel time channels*, now not in fiction but in reality. Different countries will live in different historical epochs, their social clocks running at different rates. In the use of technology, a person may jump to the twenty first century, but psychologically may still live in the old twentieth century. If we forget about this social 'twin paradox' we may get into trouble.

(2) The Earth Summit in Rio de Janeiro in 1992 concluded that bio-diversity is essential for the survival of the biosphere under changing climatic conditions. In the same way we must expect that the future will be *pluralistic*: instead of a single political–technological–economic dogma,

different models will be used in parallel on different continents, under different cultural climates.

(3) We need to respect the *complementarity of models*, all containing some elements of truth in spite of the fact that they contradict one another, not only as in quantum mechanics but also in the world in which we live. Our minds will need widening to embrace *tolerance*. As Pope John Paul II said in Assisi: *The acceptance of being different is the only passable way in Europe, if we intend to be worthy of her cultural heritage* (9 January 1993). Or, as Niels Bohr once put it: *Physicists have learned that concepts should be treated with extreme care if we are going to use them beyond our own everyday experiences. This indicates the dangers we are confronted with when we are judging alien cultures from our personal point of view.*

(4) *Information* will become more valuable in the market than steel and energy. Japanese home electronics have sold better because they are smaller and smarter. This makes a great difference: strict conservation laws are valid for materials and energy, but information can grow. Here is good news for poor nations!

(5) In place of national borders we shall have to think in *global perspectives*, because none of ballistic missiles, the supply of gasoline, world trade and mass migration, sky TV and jet tourism, the AIDS epidemic and air pollution, or climatic change and the ozone hole, respect such boundaries. National borders will exist mainly in the memoirs of retired generals.

(6) Education should not try to fight the information explosion. This is a problem only for senior citizens. For a new-born baby everything is fresh and beautiful. The duty of the school will not be to instruct in *correct knowledge*, but to *preserve our childish creativity*.

(7) Science teaching has also to contribute to moral education. Modern sexual behaviour is more influenced by the Pill or by AIDS than by any printed moral code. The population explosion cannot be controlled without scientific literacy for girls. Freon molecules last for a century; greenhouse warming and the ozone hole follow CO_2 and CF_2Cl_2 emissions after a delay of decades; the latency of lung and skin cancers exceeds a decade. *The sins of the fathers will be visited upon the sons.* Global problems may be irrelevant for politicians, who look only towards the next election, or to industrialists who look only for a rapid return on investment; they are relevant mainly for our children and for our pupils, who will inherit this planet from us. This is why science is going to be one of the most important school subjects in the future. This goal has been set by the UNESCO-sponsored international Project 2000+.

Developing nations which still live in the agricultural era, trying hard to feed an exploding population and losing brain power, intend to catch up. How can they do it?

Should they move towards a classical industrial revolution (towards the London of the 18th century or the Chicago of the 19th century)?

Should they introduce classical schooling, asking Northern experts to send curricula, textbooks and exam tests from Cambridge, Harvard or the Sorbonne, because their local professorial elite was educated there 50 long years ago?

Following these paths, they would invite pollution, drugs, unemployment and social troubles, meeting in the twenty first century the problems the North already faces today. The gap between haves and have-nots would then only become wider.

Or can they find a *shortcut to the twenty first century*, in order to arrive there on time? Is it possible to jump straight into the post-industrial era
- by having an appropriate educational style?
- by educating in the approach of modern science?
- by encouraging open-ended creativity?
- by learning to enjoy the era of information (with its video-games and e-mail)?
- by thinking and acting without selfishness, with global responsibility?

Of course, these are generalities. But what should we teach in the physics lesson in the classroom? What should we teach instead of mechanics, electricity etc? Go and ask an active research physicist. He or she will tell you that:
- the message of *mechanics* is that the state of a thing is conserved if it is left alone (inertia), but that it can be changed under the influences of other partners in the environment (interaction); by understanding *laws of motion* one can *anticipate events and create technology*;
- the message of *thermodynamics* is that the world is made of atoms, thus it has many degrees of freedom; that in such a world disorder emerges from order, increasing spontaneously, and that one can create local order according to plan only by making a larger amount of disorder (pollution) elsewhere; that this is how life and technology work; that this is why we consume food and fuel;
- the message of *electricity* is that with the interplay of electrons and fields one can handle energy and information; that these can make humans great; that even present politics, trade and music is done by electronics;
- the message of *atomic physics* is that our traditional concepts, and indeed our brains, have their limitations; that to penetrate into the structure of matter we have to use models as grades of understanding; that we have to live with complementarily contradicting models and to respect them; that such tolerance will bring rewards (both in high technology and in politics);
- the message of *nuclear physics* is that it is important to be able to deal with the duality of opportunity and risk; of power and responsibility.

These are concepts which are valuable even for would-be poets and politicians. As our teachers used to say, we have to keep in mind that the prime ministers of the twenty first century are schoolchildren today, and we teachers have a chance now to speak to them and to educate them.

But how do we teach the art of model making instead of that of memorizing rules?

Every baby has to survive the first hard years of life in a strange unknown world, so every child is a born modeller: each toy and each game is a model of the grown-up world. They do not mistake a doll for a baby or a Matchbox car for a Mercedes, but just enjoy using them. Verbalization, reading, writing and reproduction are much harder jobs needing to be learned from adults. Skills taught in a strange language are harder for Third World children than for TV educated Northern city children. If we offer the scientific— trial and error—approach instead of a collection of data and formulae to be accepted and memorized, starting in the primary school, then the Third World may even have an advantage! *Nearest to the genius is the child*, said the Hungarian–American mathematician, Cornelius Lanczos.

One of the basic concepts in science (and in society) is *interaction*. In grade 3 it just means the collision of marbles (momentum transfer). In grade 10 it means action at a distance (force). In grade 11 it means fields (electricity, magnetism, light, radio waves). At college it may be explained in terms of emission and re-absorption of quanta. *Interaction is always mutual*, with both partners being affected, as is expressed in Newton's Third Law (the principle of action and reaction). It is very exceptional for one of the partners not to mind it very much: one-way-action (the one-body problem) is a model of very limited validity, restricted to cases of a very massive partner. This is why the conceptual separation of the *observer* (who obtains information) and the *object* (which is observed but not disturbed) may serve for a while to emphasize the objectivity of science, but whose limitations show themselves in quantum mechanics and in psychology. All the classical concepts from one-body problems—collision, force, field—are *transient* but *valuable models* of interaction, as will be those new concepts yet to be discovered in the twenty first century.

One of the oldest and yet also one of the youngest concepts of science is the *atom*. In the early grades molecules are just rigid marbles (the kinetic gas model), then they are sticky balls (capillarity). Later on we can explore their funny shapes (differing specific heats of gases), and even their size (the thickness of a layer of petrol on water). If you have got enough energy, you can break molecules (electrolysis). At first sight, atoms are rigid building blocks with definite valence-hooks (chemistry), but then we can tear electrons off them (photo-ionization). We can depict the composite atom first as a tiny planetary system (Rutherford), then as standing wave patterns (de Broglie). Finally, before leaving high school, you may explore the reaches of your imagination by discussing particle–wave complementarity.

In the last year of his long life, Niels Bohr tried to explain the power of complementarity to the professors of humanities at the University of

Copenhagen.   He asked them, *What could be a complementary concept to justice, in such a way that it occasionally contradicts, but is still to be respected as of equal value?* Later, he answered his own question: *Love*.

> *Travellers! Here there is no path.*
> *Paths are made by walking.*
> *When you look over your shoulder,*
> *You see a path you'll never walk again.*
>
> *Antonio Machado*

(Supported by the Innovation of the Pedagogical Profession Fund)

In looking to the future, we see a wide field of facts, laws, theory, speculation—scientific knowledge—that will grow in extent and, we hope, be reduced in complexity. This study of nature that is a part of man's intellectual life is a delight to all scientists. In their hands, science is not just a business of collecting facts or stating laws or directing experiments. It is above all an <u>art</u> of sensing the best choice of view or the most fruitful line of investigation for a growing understanding of nature....

The future of science, both as practical knowledge and as an intellectual heritage, depends greatly on the attitude of laymen: parents, teachers, administrators, governors, ... all educated people; so it is in your hands, as a member of a scientific civilization, to maintain the good name of science.

Eric Rogers *Physics for the Inquiring Mind* pp 757–759

CHAPTER 2

A VULGAR SCIENCE CURRICULUM

Jon Ogborn

Adapted from a seminar given at the Centro de Investigacion y Desarrollo de la Educacion, Santiago, Chile, June 1991.

2.1. VULGARIZATION

I see no escape from the conclusion that a science curriculum intended for everyone must be vulgar. Latin languages speak of 'vulgarization' where English speaks of 'popularization'. In a sense, the conclusion is simply a tautology. Science for all the people must be fit for all the people. This is what vulgarization *means*. But of course it says more than that. It says that a science for all the people must be attractive and interesting, and should show a reason and purpose for science. Its function is not to make all the people able to do science. Doing science is a valuable, specialized, and nowadays 'industrialized' activity. It is important and exciting, but not the most wonderful activity in the world, as we scientists sometimes pretend. Just because it is something done by a few, and because what it does affects us all, it matters that people at large have an idea of what it can and cannot do.

2.2. THE STORIES OF SCIENCE

Scientific knowledge consists of world-stories, of tales of the unknown hidden inside or behind the known. These stories evoke the world-pictures which science has made. And these stories answer questions, such as 'What is life?', which are important to people.

I think we can, for the science curriculum, usefully divide the stories in which scientists have something important and interesting to say, into five kinds. It is important that there should not be too many kinds. We need as simple a picture of the shape of scientific knowledge as we can have. The reason is that when a child goes home to its parents, and they ask, 'What did you learn today?', the child should have an answer to give. It is equally important that the school can tell the parents what is happening. Only if the scientific content is built around a few simple, important themes is this possible. If not, science in the school becomes something the parents feel they know nothing about.

My five themes are:

Life
Matter
The Universe
The Made World
Information.

You notice that these themes do not mention the name of any science subject. The important questions of science are not the exclusive property of any one subject. The question, 'What is life?', was answered partly from biology, partly from biochemistry, and partly from X-ray studies from physics. All of them collaborated to find an answer.

This means that if you approach the content of the curriculum in this way, you have two choices. You can choose to teach science organized by subjects, but referring all the time to such broader themes. Alternatively, you can combine and teach science as one subject, organized by such themes. Whichever one does, I believe it to be essential that the child and the parents can see that what is being done relates to important human questions such as these themes express.

There is another traditional way of finding themes. You take a topic which is rich in possibilities, for example, 'Water'. Now, it is quite true that if you investigate 'Water' you can invoke many interesting ideas from the sciences. But if you say to parents, 'We have a good science curriculum; we teach your children valuable things; we tell them about water', the parents will not understand. If you say to them, 'Water is our excuse, our reason, our beginning point', they will understand, but will then want to know where you are going from that beginning. Eric Rogers taught us the importance of clear 'end-points' in the curriculum. What I want to insist on is that these end points should relate to essential human questions, in which all people— children and parents—have some interest. If anything I have to say reminds you of Eric Rogers' thinking, then it is either because I stole it from him or because he inspired the thought in me.

I will now say something about each of these themes. What are their essentials? What I will discuss is not yet a programme of teaching: we are establishing the goals, the end, not the beginning and not the path.

2.3. THE NATURE OF LIFE

The first part of an answer to the question, 'What is life?', is a strange equation:

evolution = replication plus competition.

If organisms reproduce themselves, then any which gain a small advantage and reproduce more quickly, rather quickly displace other varieties. So we get a branching tree, and life takes an enormous variety of forms, changing all the time. The message is that evolution equals photocopying plus fighting. This system is however not stable. A small change in the environment tends to induce qualitative changes in organisms. It can destroy a whole species, but it can cause a new branch of life to develop. It is for this reason that we see such a huge variety of organisms.

The motor of all this is not the survival of the individual, nor of the species, but is the survival of genes. An individual can certainly try to survive. We can (wrongly) project this wish on to a species. But genes are just segments of DNA code. Genes have no wishes, they intend nothing; all they do is replicate. 'Success' is just to be there in as large a number as possible. To succeed means there are many of them; that there are many of them means they succeeded.

Further, these genes are all the same kind of entity. All life has a common basis. Everything that we know, leaves, plants, microbes, animals, our cells, uses exactly the same fundamental mechanism of replication: the DNA molecule. At the molecular level, life is very uniform, despite its extraordinary variety as we perceive it. And this has to do with life understood as information; many messages can be written in a small alphabet.

A different aspect of the question, 'What is life?', is that complex organisms work by operating as entire systems. Each part works with the others to the benefit of the whole. There are many ways to do it. In the case of humans and many of the animals, we have developed internal organs; heart, liver, kidneys. Other creatures manage it in different ways. But in each case each part has a function in relation to the whole. And so a lot of biology consists of describing how these systems operate together. We might say the story of life is two stories: replication and competition, and co-operation.

I think that we should also relate this fact to a different theme, that of the made world. When we make things, whether television sets or computers, we are also building systems made of parts which must co-operate. These two facts need to be seen together. In this sense a computer is like a person: the parts work together for the whole.

When you are a child, the world around you is the world of things on your own scale. It is very strange to you if somebody says, 'On your skin are tiny invisible bacteria'. It is difficult to believe that the clean-looking water from the tap might be dangerous. But the basic entities of life, cells and DNA molecules, are microscopic and too small to see directly. So the child has to learn that beneath the world of everyday things there is a world which is not visible to the eye, but is just as real and perhaps even more important.

When you teach chemistry or physics you find it very difficult to persuade children that under the surface of things lie atoms and molecules, too small to see. Starting in the primary school there is a ladder of size to climb down, going inside matter. It can start by collecting dirty water from a puddle, and seeing the dirt floating in it with a hand lens. Then continue with a microscope, looking at microbes. To start there may greatly help later teaching about molecules and atoms.

Looking at science as a whole, one of its essential stories is that the real world is deep inside the world we think we know as real. It is important that we can say to children, 'Things are not what they seem; do not always trust appearances'.

2.4.   THE NATURE OF MATTER

In this second theme, the nature of matter, we also move progressively deeper and deeper inside matter. My picture (figure 2.1) illustrates the idea. The picture begins with the ancient Greek categories: Earth, water, air and fire.

Earth stands for solid things. There are several important kinds of solid material. Strangely, one kind that is all around us is almost never spoken of in the secondary school: namely brick and glass, that is to say, ceramics. We drink from ceramic cups, and live in houses made of brick or stone with glass windows, and never speak of this kind of matter in science in school. Ceramics are good for making cups because they are hard and rigid, but less good because they are breakable. These properties arise because their atoms are joined strongly with chemical bonds, but in an irregular pattern.

Metals are quite different. Their atoms are arranged in a rather regular way. And they are held together not by bonds linking one atom to the next, but because the atoms live in a sea of electrically charged glue. Thus if you pull a piece of wire, you can stretch it. The atoms can slide one upon the other within the glue. This means we can shape metals. We can press sheets of steel and out comes the shape of part of a car. The movement of atoms inside is made easier by the existence of defects in their regular pattern of arrangement, and these defects (dislocations) move around rather easily. Only one particle need move to pass a defect from one row of atoms to the next. It is like moving a carpet, not by pulling the whole carpet at once, but by making a small fold and moving the fold along.

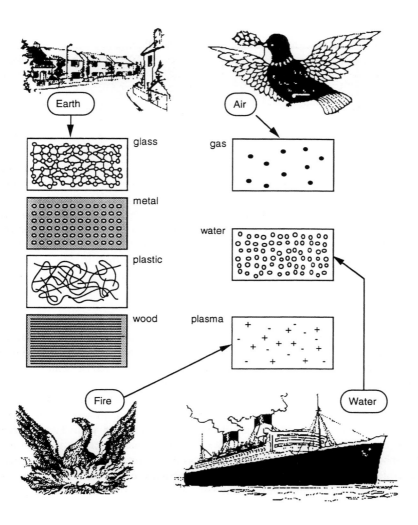

Figure 2.1. *The Nature of Matter*

Another important kind of material, also hardly spoken of in the school curriculum, is plastics. Children bring their food in a plastic box, but the existence of plastics is only admitted in chemistry lessons. These materials are made of long string-like molecules. They are essentially molecular spaghetti. This is why when you tear open a plastic bag it easily stretches and tears. Moving the molecules is like lifting spaghetti from a pan. The molecules can slide around and move easily. Nor should we forget that the proteins essential to life are polymers.

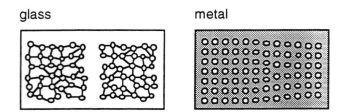

Figure 2.2. *Fracture in glass; dislocation in a metal*

We are now telling stories not just about what is inside matter, but about how it accounts for our experience in the world of these materials. So going inside is not just seeing new things, but is thinking about behaviour at the atomic level which explains behaviour at the everyday large scale level.

Another material whose existence is hardly admitted in the school is wood. Wood consists of strong fibres inside a soft material. This has a big advantage. If you pull it, then the fibres may break, but the material does not come to pieces, because the soft material can flow and hold the fibres together. Wood is a composite material which can be very light, very strong and very tough. Recently, humanity has imitated nature and manufactured other composite materials which work in the same way, fibreglass for boats and for pole-vaulting, and in some modern airplanes a composite made of fibres of carbon.

The most difficult kind of substance to understand is the simplest, the one deriving from the Greek category of air, that is, gases. When I look across the room I have no feeling of anything between me and you. There seems to be just emptiness between us. We can move about freely in it. We have no sense that the air in the room is somehow a material equally as real as the table or ourselves. We can perhaps persuade ourselves. We wave our hand, we feel something flow around it. Also, the wind blows down a tree in a storm. That seems more real.

There is a very real problem of persuading children that the air is a real substance. Why is it that the air between us seems just to be empty space? How can I see you through something? How can I move about in it? All this happens because the particles of air are a long way apart, and between them there really is nothing. So most of the air is literally nothing. If you teach children about molecules in gases, and then ask them to draw a picture of the nature of a gas, they will obey you politely and draw particles with gaps between them. But if you ask them what is between the particles, they say, 'Gas, of course. What else could it be?'.

So this story begins to show how science has a function in stretching the scientific mind and imagination. Matter, which seems so obvious and rather

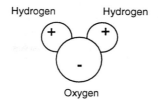

Figure 2.3. *The water molecule*

dull, is full of surprises.

If we think about fire, then we get another level of surprises. It was a very good choice of the Greeks to make fire the fourth element, because there really are four kinds of substances: solid, liquid, gas and plasma. A plasma is what you get when atoms are pulled apart into electrons and ions separately. The Sun is made of a plasma. So is a flame. Because a plasma contains freely moving charged particles, the flame of a match or a candle conducts electricity. The atoms are broken into charged pieces and these charged particles travel in the flame. Metals are also materials where charged particles, the electrons which also glue the metal together, are able to move. So metals usually conduct electricity.

Another very interesting case is water. Water molecules are boomerang-shaped, with a hydrogen atom at each tip and the oxygen atom in the middle. As a result, the molecules have an electric charge on the two ends. This is why it is easy to dissolve things in water. If you put salt into water, then the sodium and chlorine ions collect polarized water molecules around them. If water did not dissolve salts easily we could not run, jump or think, because our nerve impulses come from little electrical batteries driven by differences in concentration of salt inside and outside the nerve cell. Our thoughts depend on the angular shape of the molecule of water.

Quickly this story descends to a deeper level, and we have to speak of the structure of atoms, with electrons surrounding a nucleus, again with nothing in the space between them. Then we go inside the nucleus, itself made of new particles, protons and neutrons. Finally we can even go inside these, finding them again made of new things only discovered in the last twenty years.

So, we have a story of successively deeper and deeper structure. I am not saying that we can reach the end of this story in the secondary school. I think not. But I do think the curriculum should have this story in its mind so as to present to children the possibility of always going a level deeper.

2.5. OUR PLACE IN THE UNIVERSE

The next theme is the nature of the whole universe. The present opinion is that the universe began with a large hot explosion called the 'Big Bang'. In this explosion there was time to make hydrogen, but not enough time to convert more than about ten percent of that hydrogen into helium, so that most of the universe is hydrogen. There was no time to make any appreciable amount of any heavier elements. If there had been enough time, the whole universe would be made of iron, the most stable nucleus. The 'Big Bang' went quickly.

But this makes a problem: where do we come from? We are not made entirely of hydrogen. Essential to our life are metals like iron for the blood, phosphorus for DNA, carbon for organic molecules. Where did they come from? We now think they came from the supernova explosions of early, very big stars. Our material was manufactured inside a star in a huge explosion. So we might say that we are the ashes of exploding stars. All the important elements were fabricated not at the beginning but after the beginning. Some of this ash condensed into planets, on one of which we happen to live. Having such a planet sitting near a burning star it became possible, after a long time, that life could evolve, so that here we are to try to understand the story.

We have to teach about the universe if Newtonian mechanics is ever to make any sense. The universe is made of matter in ceaseless motion, motion which needs nothing to keep it moving. So when we state Newton's first law, as we should, in the form, 'Motion has no cause', it is only the ceaseless motion in the sky which makes such a statement other than ridiculously contrary to obvious experience.

2.6. THE MADE WORLD

The themes, Life, Matter and the Universe, cover much of the scientific picture of the natural world. However, science is used not just to understand the world, but to change it. So we need the theme of the made world, the artificial world and the world made by people. Here I think there are essentially three aspects to consider: our acting upon the environment, our use of materials, and our uses of energy.

When we talk about technology we often think of modern high technology: airplanes, fax machines, motorways, or computers. But I suppose that really the most important technologies are much simpler: houses, clothes and agriculture. Thus there is a lot to be said about the made world quite early, much of it in the primary school, about the early technologies of human beings, in which they first began to contrive artificial worlds in which to live. There is, of course, much to say here also about the social organization of human beings. You cannot farm or live in houses if you do not have some social organization. Farming and towns induce social structures. The beginning of talking about the artificial, technical world could be talking about ancient civilizations.

With materials and with fire, people began to make ceramics and to extract metals. Pots were fired from clay and then, by accident, someone found a little pool of metal in the fire. The use of fire was an important beginning of our technologies.

Then we had to devise ways of obtaining clean water. With small populations, it is not a big problem, but as civilizations grew and cities were built and had dense populations, providing clean water became a major problem.

In the Western countries coal and iron were exploited to make steam and steam engines, and this began to drive a whole new development of social organization through the use of machines. We built ourselves factories, and in the process made another social structure based on differences between people whose wealth was their labour and people whose wealth was money. As well as factories, we made cathedrals reflecting systems of beliefs. The manufactured world mirrors and transforms the social.

I believe that such stories are an essential part of a science curriculum. But they are a part where the science curriculum touches others. These stories cannot be told without help from teachers of history and sociology. They should not be told without speaking with teachers of literature. So it is a theme in which science is not autonomous. But it is also a place where science has something useful to say and some important stories to tell, which are of interest to other parts of the curriculum.

Consider for a moment the case of electricity. How hard it is now to imagine the world without electricity. Perhaps every school in an industrial country should turn off the electricity for a whole day once a year, and try living without it, to realize the difference it makes if one is obliged to stop work or play when the Sun goes down, unless one has a candle. And one cannot telephone, power machines, or stay warm without a fire. This is what it was like only a few generations ago, and still is for many. Yet this development has led to telecommunication, satellites and computers, besides electric light and electric power. We have become an electricity civilization.

2.7.   INFORMATION

The fourth theme is a modern one. We are not only an electricity civilization, but also an information civilization. We have telephone, fax, radio, television, computers, telecommunication. These are now beginning to affect the social structure, but in ways we cannot yet foresee. The children to whom we are teaching science will live in a time when the social structure has been re-shaped through information technologies.

Information has a scientific dimension as well as a technological one. Information is not just words on paper. It is also an abstract idea, and in its deepest sense it is again the root of life. The DNA molecule operates by being a coded message, a fax to the cell. It has a code which tells the cell which proteins to make, and in the end that message decides how the cell develops into a part of one organ or another.

There are concepts associated with information which are essential to living in the modern world. They are concepts about systems, because a system is fundamentally controlled by flows of information. The main ones are the concepts of feedback and control. In your car, you press on the pedal and the car goes faster. In this case a wire carries the information. But control is needed just as much for your own body to function. Your body needs to stay at a constant temperature, so if you become a little too warm you perspire and cool down again. Your body is continually maintaining a steady state by using a flow of information to tell it what to do. We should have children look at their hands with a lens and see the water come out of the pores. Why should it come out? Information flows from the temperature of the hand, sending nerve signals to open the pores. The concept of feedback, a looping of information going around and around, is essential. It can keep things stable, or it can produce instability or oscillation.

Information can also contain errors. Errors in the DNA code are part of the driving force of evolution, with the possibility of a mutation leading to the possibility of a new adaptation, even a new species.

It is also necessary to protect against errors. In a computer, if you think of it, it is very surprising that so few errors occur. You have some millions of pieces of information each second coming to the screen, and it is really very rare that there is a mistake. This is not at all because there are no mistakes. It is because the mistakes are located and corrected. This is often done by clever methods of coding the information, so that a message contains itself more than one time, and the two copies can be compared.

Thus there is a whole story to tell about the scientific and mathematical aspects of information, ideas which are going to be very important in the next fifty years in order to understand the impact of information on society. And they are also ideas which are essential to understanding ourselves as biological organisms. So here are two good reasons why information is a theme for the science curriculum.

2.8.   STRANGE NEW ENTITIES

The stories of science populate the world with a zoo of strange new entities. Let us take one example, the idea of a field.

Is it not a strange notion, that the Moon up there makes the tide in the ocean down here rise? Humanity seems unable to bear the idea that one thing can act on another without touching it. So empty space was filled with something called a field, for the Moon to be able to 'touch' the ocean.

Empty space gets a number attached to every point. This crazy idea turned out to have tremendous power. When you watch television you are watching pictures whose information was carried by such fields, which not only consist of numbers attached to space, but which have also grown legs. The abstract concept gets up and walks. At first Newton was rather embarrassed by his invention of the gravitational field. But after many years, others invented electric and magnetic fields, Faraday took them seriously as real entities, and Maxwell proved that they could travel. So we have the story of television, communication, information—and of light.

This is a story of the seeming madness of science. I said before that it is very hard to persuade children that there is a microscopic world within our daily world. It is even harder to persuade them that invisible, intangible things are real.

2.9.   THE NEED TO BE VULGAR

We usually treat scientific knowledge as a kind of high technical knowledge which can only be acquired very, very slowly.   But good television programmes about science feel free to tackle big questions without insisting on years of previous preparation. We need to make science lessons more like that. There are many good examples to go by, including popular magazines and good active science museums. And I do *not* mean presenting old bits of science curriculum in bright new cartoons, as most textbooks do today. I mean going for big important questions. Why else should anyone listen?

Students need ideas—not a single view but discussions of views—concerning: laws in science, the relationship between experiment and theory, and the strong distinction between theory and simple hypotheses. For these benefits students must be carried through actual examples, not just harangued or given pat definitions. Hence my choice of history of astronomy (for growth of deductive theory); and magnetism (to show theory yielding language); and kinetic theory of gases (for rich predictions and promoting of understanding).

Eric Rogers *Teaching Physics for the Inquiring Mind* p 30

CHAPTER 3

TEACHING ABOUT THE NATURE OF SCIENCE THROUGH HISTORY

Joan Solomon

For Eric Rogers, discussion of the nature of science was an essential part of any science course, and he saw that this could sometimes be done through the history of science. The 1989 version of the National Curriculum in Science for England and Wales contained a section on the nature of science, including some history of science. Subsequent revisions moved this work into a section on scientific investigations and although it now has a lower profile than at the time the work described here began, it has not disappeared completely.

The introduction to the curriculum stated:

Pupils should develop their knowledge and understanding of the ways in which scientific ideas change through time and how the nature of these ideas and the uses to which they are put are affected by social, moral, spiritual and cultural contexts in which they are developed.

The British school system was largely unprepared for this innovation. Not only were there almost no classroom resources, but it was also uncertain what the effects of teaching it might be. Worse still, it was not even clear precisely what the defects were in the pupils' understanding of the nature of science that this addition to the curriculum was supposed to remedy. Thus an aim of our work was to observe and record how learning science through historical studies might affect pupils' understanding of the nature of science.

3.1. THE PLACE OF HISTORY OF SCIENCE

The general value of an historical approach to science teaching has been argued for well over a century. Jenkins (1989) traces it back to 1850. In the last twenty years the list of educators from many countries who have recommended a study of the history of science in high schools has become too long for us to catalogue here (see for example Brush 1974, Dana 1990, and Duschl 1985). Possible benefits commonly claimed include:

- a better learning of the concepts of science;
- increased interest and motivation;
- an introduction to the philosophy of science;
- a better attitude of the public towards science;
- an understanding of the social relations of science.

Amongst those who have contributed to this discussion have been Aikenhead 1979 (social aspects of science), Brouwer and Singh 1983 (motivation), Sherratt 1982 (humanistic and cultural), Matthews 1989 (philosophy), Wandersee 1986 (vignettes and scientists), and Winchester 1989 (context of discovery). Gruender and Tobin (1991) have also argued a case for illuminating the psychology of scientific discovery:

> ... *the interplay of creative, improvisational and disciplined efforts, along with the alternation of excitement and elation with disappointment which lie at the heart of scientific inquiry.*

Those who espouse a relativistic approach to the philosophy of science have been less likely to recommend study of the history of science in school (e.g. Collins and Shaplin 1989, Millar 1990). Others like Hodson (1986) and Selley (1989) have pointed out how pupils with a naive approach to 'correctness' in science might find it difficult to benefit from historical study. Many however have commended it strongly for the pre- and in-service education of teachers (e.g. Ledermann and Zeider 1987 and Tamir 1989) where it may serve to combat naive empiricism.

It is not unusual to find Piaget's early guess that young children's thinking might recapitulate early concepts in the history of science also used to promote an historical approach to teaching science. However studies of children's so-called Aristotelian views show, not surprisingly, that, as coherent conceptual systems, the thinking of ancient philosophers and of untutored children have very little in common.

Writing about the place of history of science in science education has usually been theoretical, arguing the case for its inclusion rather than documenting how this could be done. Notable exceptions include Klopfer (1969), Rutherford *et al* (1970), Pumfrey (1989) and Johnson (1990). Even here, however, there has been little close exploration of the growth of pupils' understanding while the historical teaching was carried out, or of the summative effects of such a course. It was such an empirical study of pupils' progress which was the main focus of our research.

3.2.   THE CLASSROOM MATERIALS

In preparing materials for teaching about the nature of science, we need to bear in mind that the aims discussed above nowhere include an introduction to the history of science for its own sake as an academic discipline. This gives us some latitude in historical precision should it conflict with other objectives. Kuhn's well known Principle of Incommensurability warns of the enormous hurdles in the way of those who try to understand earlier modes of explanation, whether adults or children. In our research, extracts from history have not only been chosen for their ease of comprehension, and for their fit with syllabus content, but have also been carefully selected to emphasize those points about the nature of science which preliminary findings showed that pupils most needed to learn.

Hence we chose to show social context and the emergence of new concepts, rather than analyses of ancient perspectives or exact chronological progression.

Teaching materials were prepared and used in the main classroom research, in three different schools. They have subsequently been published as *Exploring the Nature of Science* (Solomon 1991). Without this common resource it would have been impossible to combine data from different classrooms in a meaningful way.

We used topics which would need to be covered at this stage in the national curriculum, and taught them in an historical context. We also included practical laboratory investigations whenever possible. The following brief outline of the unit *Mountains on the Moon* may give some idea of the character of our materials:

The story of how the telescope was discovered and then used by Galileo is set out in eight short sections. The first tells how lenses were first used by medieval monks for reading, goes on through the Dutch discovery of the telescope to Galileo's improvement of its magnification, and his recommendation of it to the ruler of Venice for military use. The later sections cover how he saw shadows on the Moon growing longer night after night, interpreted then as evidence for mountains and calculated the heights of these. The older theory of a smooth Moon shining with its own light is simply related to contemporary belief in heavenly perfection.

The class reads the whole story, and then each group of pupils makes a poster to show the content of their section of it. These are exhibited around the classroom. A model of the Moon is made from a football embellished with mountains and craters made from modelling wax. It is illuminated from one side by a projector beam, and observed by the pupils.

Finally the pupils design and carry out their own practical investigations showing how the measured lengths of shadows are affected by (a) the height of the object and (b) the inclination of the light.

[In this way national curriculum requirements under skills of exploration, optics, and astronomy are covered at the same time as statements of attainment for the nature of science.]

The units were not all designed in exactly this format, but whenever a short account of a piece of history was given, in written, cartoon or other form, a short activity was set for the pupils. This could be a DART (Directed Activity Related to Text) such as making a poster, sequencing a set of statements, running an experiment or taking part in a role-play. Its purpose was to encourage pupils to re-examine the 'text' in order to extract as much information from it as they could.

3.3. PILOT WORK AND THE CONSTRUCTION OF A THEORETICAL PERSPECTIVE

We were fortunate in having access to data from two investigations into pupils' perceptions of scientists and their work. The first followed in the tradition of the classic work by Mead and Metrau (1957) drawing a picture of a scientist, but in addition interviewed pupils in groups about these pictures. Pictures mostly showed the usual balding males, often bandaged from the results of previous experiments, performing chemical operations. During interviews, however, these young pupils also insisted that

- scientists were 'not really like that', but that is how you draw them;
- scientists probably look like everyone else;
- scientists often 'explain about the ozone layer (or other topic) on the television'.

The second line of research involved free writing in response to questions about why scientists did experiments, and whether they knew what they expected to happen before they did an experiment. Frequent negative responses to the second question demonstrated a 'shot in the dark' attitude towards experimental investigation more extreme, it seemed, than simple empiricism. This could be called *'serendipitous empiricism'*. From this preliminary work and results of a questionnaire given to 400 pupils, we knew that there were certain features of pupils' perceptions that we should anticipate:

- pupils' *serendipitous empiricism*;
- the different and varied meanings pupils hold for 'theory' (fact, guess, prediction, and explanation).

In addition the literature about teaching history, and our own experience, suggested two more points to which we should attend:

- a 'Whig view' of history, together with a naive two-value system likely to produce a view that scientists' theories had simply been 'wrong' in the past (Butterfield 1931, Selley 1989);
- difficulty in empathizing with the social and cultural attitudes of a previous age.

3.4.   METHODOLOGY FOR THE CLASSROOM RESEARCH

The effect of teachers' views about science on the perceptions of their students has been interestingly documented by Brickhouse (1989), indicating that it would be important to include teachers as co-researchers. We were also anxious not to rely on just one module about the nature of science (e.g. Carey *et al* 1990) but to build upon work used at regular intervals in the normal teaching programme over a whole year. We operated in five classrooms located in three schools situated in very different geographical locations. In each of these classrooms a researcher worked alongside the teacher on a regular basis. We used elements of three kinds of research methodology:

- an intervention study, where new material was introduced and its effect monitored;
- action research, where we were partial observers trying, alongside the teachers, to recognize and bring about good practice;
- experimental research where impartial observation and measurement were designed to probe and explain progress in pupil understanding.

There were constraints in combining these procedures. In intervention research it is usual to monitor all the actors involved, but because this was also action research, our partnership with the teachers precluded judgmental observation of their work. On the contrary we took careful note of all their perceptions and recommendations. Our preliminary questionnaire data was more in the tradition of objective experimentation than action research, but we were able to amplify the responses through a succession of interviews.

Teachers were allowed free choice of any six from the thirteen units so that these could be fitted into the scheme of work with the least possible disruption. (Two additional units were specially written for one of the schools which had an unusual scheme of work.) Units contained alternative DARTs from which teachers chose the one they judged to be most appropriate. The research nevertheless had theoretical coherence, resulting from a shared understanding of the pupils' problems which grew out of our preliminary studies. This common perspective, together with continuous contact and discussion between teachers and researchers, the researchers' own experience of classroom teaching, and above all the enthusiasm of the team of teachers, helped to make this diversity a source of richness rather than of incoherence.

3.5.   TALKING ABOUT EXPERIMENT AND THEORY

Before the course began, pupils were interviewed about their answers to the questionnaire. It was possible to find pupils who expressed a single coherent view about the roles of experiment and theory, and others who flitted from one image of science to another. The first example is from an 11 year old boy, Leo, who holds a typical cartoon view of experiment. He said firmly about expectations:

You don't know what is going to happen, like if it's going to spit or explode or something. Or stay the same.

Leo gave 'heating crystals' as his example of an experiment. He also put down that theories were 'facts', and maintained this cartoon view of science unchanged throughout the interview.

Sab, a girl in the same class, produced mixed responses as she moved from one domain to another. She wrote that scientists do experiments to try out their explanations for why things happen, and then later that they did NOT know what they expected to happen before they did the experiment. She also wrote that theories are facts, not explanations, although she was 'not sure about this'. On being questioned she started by only saying 'I don't know' two or three times. Initially her friend Bet had kept quiet.

Int: So what sort of experiment would you think about, would be like that?

Sab: Well like Mr R (the teacher). He's done it before, he shows all of us.

Bet: He knows.

Sab: He's done it before.

Bet: He's done it with other years.

Int: And if it was the first time?

Bet: I don't think he would know.

Int: Now look you put down 'explanation' there. You said,'Scientists do experiments to try out their explanations for why things happen.'

Sab: Yea, yea it is (emphatically).

Int: But you didn't think a scientific theory was explanation, but facts.

Bet: I reckon it's the facts.

Sab: I was going ... I thought of putting that one ('explanation').

Int: Do you use this word 'theory' much, or not much?

Sab: Not much.

Int: So perhaps you were just guessing?

Sab: Yea.

Bet: We don't even use it at all.

Both: No ... just 'experiments'.

This transcript suggests that there are three substantial hurdles for such pupils. In the first place they have access to a number of different co-existing meanings and images. Secondly they have little or no knowledge about scientists, past or present. Thirdly they are unfamiliar with the word 'theory' and, as their questionnaire responses showed, they know no examples of scientific theory. All these problems were addressed by the intervention materials.

After the course, the populist images of scientists still existed, but by now pupils also have access to stories about real scientists, their theories and experiments. Two low ability 11–12 year old boys are speaking after the course about what a theory is.

Avl: It's when they (the scientists) think something is ... When they don't know for sure but they think that's what the answer is.

Int: Can you tell me what a theory is? Is it a guess?

Avl: Yea, it's a sort of a guess.

Int: Why do they make the guess?

Dav: Because they sort of...

Avl: Because if they don't make a guess they won't have a clue what's going to go on, will they?

Dav: Yea, that's what I was going to say.

Int: Who was it that found out about smallpox?

Both: Jenner.

Int: What was his theory, do you remember?

Dav: If you give someone cowpox and they catch smallpox they wouldn't die.

Int: OK. So how did he get his theory? Was it just a guess?

Avl: No. It was because the milkmaids never got the pox.

In the following extract the boy, Leo, who had so strong a cartoon image at the beginning, is talking with his friend at the end of the course. Like the boys above, they need to be triggered to remember instances of theories they have learnt—the particle theory, or Galileo's theory about the Moon.

Int: Now can you tell me what a theory is?

Leo: It's like you are guessing about something. You think.

Int: Are 'guessing' and 'thinking' different words?

Leo: Yea.

Int: So why did you use the word 'think'?

Leo: Because he thought.

Leo seems to have moved away from 'facts' and nearer to an 'idea'.

Other units in the course took up the idea of scientific controversy. The stories were carefully chosen, like the controversy between Volta and Galvani, to illustrate different lines of investigation rather than confrontation and falsification. Towards the end of the course, some of the 13 year old pupils had clearly begun to form an idea of how the process of theory building is carried out.

Int: I just wanted some of your ideas about how scientists develop theories, how they make theories.

Rob: They study lots of things, the ways it could be, and then make up their own ideas of the way that they chose that it could be. So it is their own idea of what it is.

Int: What do they do then when they have a theory?

Rob: They can like discuss it with other scientists and see what they come up with. And keep on testing it in different ways to see what answers they get.

It was easy to find dismissive attitudes towards early theories in the children's comments. About the pre-Galilean idea that the Moon was smooth, a 12 year old girl remarked:

They didn't have the, sort of like, logic... They were just fantasizing it.

An older boy who had just come across Leslie's model of heat transfer by 'heat particles', tries to be more understanding but only manages to sound condescending.

Mar: Well I think that the idea is good for the time when he did it.

Int: For the time?

Mar: For the time and age when he did the experiment. It was three hundred years ago wasn't it?

This attitude was often supported by reference to our superior instruments. 'Technology' in several of their comments seemed synonymous with progress: *Nowadays we can see the particles (of glass) with our technology.* In the kindly condescending tone above another pupil remarked:

Well it just shows that even in those days they still tried to work out different theories. Even though they didn't have special equipment, they still had their own theories.

Such talk gives evidence of the difficulty pupils had in empathizing with the thinking of scientists whose theories they knew to have been superseded.

3.6.   DATA FROM THE QUESTIONNAIRE

Results from three questions in the pre- and post-course test are shown in figure 3.1.

The main results, all corroborated by interview data, may be summarized as:

- there is a significant movement away from seeing the purpose of experiment as making a discovery, and towards seeing it for trying out explanations;
- significantly more pupils now think that scientists know what they expect to happen in an experiment than did before the course;
- far fewer now think that a theory is a fact, with both 'an idea' and 'an explanation' being more favoured answers.

A fourth question, to which responses were surprising, was as follows.

Sometimes in the past, groups of scientists have held different theories. Is this because

   (a) they have done different experiments,

   (b) one group was wrong and the other group was right,

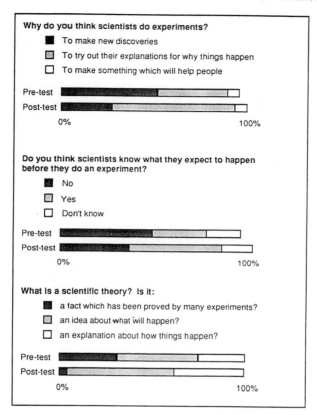

Figure 3.1. *Percentages giving answers shown to three questions, pre-course and post-course (variations in totals from 100% are due to not all questions being answered)*

(c) they looked at results of their experiments in different ways,
(d) one group made a mistake in their experiments?

Three-quarters chose response (c) *they looked at results of their experiments in different ways*, equally before and after the course. Less than a fifth chose response (a), again equally on both occasions, leaving only a handful giving the dismissive right/wrong replies such as (b) and (d), which the literature on pupils' understanding of history of science (e.g. Selley 1989) might lead one to expect. The negligible change after a year of teaching was striking.

3.7. CONCLUSIONS

There was substantial evidence that our units for teaching the history of science within the normal school curriculum contributed valuably to the pupils' understanding of the nature of science. In particular there is a significant move away from serendipitous empiricism and towards an appreciation of the interactive nature of experiment and theory. Experiments come to be seen to be designed for trying out explanations, and hence with an expectation about what may happen. Theories are now significantly less likely to be seen as just 'facts' but, like experiments, to be related to explanation or prediction.

The teachers all agreed that their pupils had learnt some of the scientific concepts better through studying them in the controversial situations in which they first arose. No doubt a part of this improvement is due to enthusiasm surrounding an innovation. However we did find evidence that helping pupils to focus on reasons for accepting one theory rather than another was more effective than just teaching accepted theory. Using these historical materials produced more durable learning.

Finally we acquired new and unexpected evidence from interviews that studying a change in theory in time past may make the process of the pupils' own conceptual change a little easier. This point is illustrated by two girls who are talking together about the difficulty contemporary scientists had in accepting Galileo's theory.

Onek: ...before we started doing things about Galileo I didn't know it *(the Moon)* had mountains and craters on it. I just thought it was a round ball that glowed.

Int: Is it difficult to change your mind?

Kar: No.

Onek: Yes, to me.

Kar: Yes, it is actually. Yes.

Onek: It seemed weird, coz sometimes when I'm watching, looking at the Moon or something, I look at the Moon and I still can't believe it, because it still looks like a round thing that glows. It's really hard.

Such reflection on the process of trying to learn is a sure sign that substantial learning is actually taking place.

ACKNOWLEDGMENTS

A version of this article appeared in the 1992 *Journal of Research in Science Teaching* **29** (4) pp 409–421

It is a pleasure to acknowledge the help of Jon Duveen, Linda Scott and Susan McCarthy in the research described here.

REFERENCES

Aikenhead G 1979 Course evaluation II: Interpretation of student performance on evaluation tasks. *Journal of Research in Science Teaching* **11** 23–30

Brickhouse N 1989 The teaching of the philosophy of science in secondary classrooms: case studies of teachers' personal theories *International Journal of Science Education* **11** (4) 437–449

Brouwer W and Singh A 1983 The historical approach to science teaching *The Physics Teacher* **7** (5) 231–236

Brush S 1974 Should the history of science be rated X? *Science Education* **183** 1164–1172

Butterfield H 1931 *The Whig Interpretation of History* (Bell: London)

Carey S, Evans R, Honda M, Jay E and Ungar C 1990 An experiment is when you try and see if it works: a study of grade 7 students' understanding of the construction of scientific knowledge *International Journal of Science Education* **11** 514–529

Collins H and Shaplin S 1989 Experiment, science teaching, and the new history and philosophy of science *Teaching the History of Science* ed Shortland M and Warwick A (Oxford: Basil Blackwell) pp 67–79

Dana T 1990 The history and philosophy of science: what does it mean for science classrooms? *The Australian Science Teachers' Journal* **36** (1) 21–26

Davies F and Greene T 1984 *Reading for Learning in the Sciences* (Edinburgh: Oliver and Boyd)

Duschl R 1985 Science education and philosophy of science: twenty five years of mutually exclusive development *School Science and Mathematics* **85** (7) 541–555

Gruender C and Tobin K 1991 Promise and prospect *Science Education* **75** (1) 1–8

Hodson D 1986 Philosophy of science and science education *Journal of Philosophy of Education* **20** (2) 215–225

Jenkins E 1989 Why the history of science? *Teaching the History of Science* ed Shortland M and Warwick A (Edinburgh: Basil Blackwell) pp 19–29

Johnson S 1990 Using philosophy of science in curriculum development: an example from high school genetics *International Journal of Science Education* **12** (3) 297–397

Klopfer L 1969 The teaching of science and the history of science *Journal of Research in Science Teaching* **6** 87–95

Ledermann N and Zeider D 1987 Science teachers' conceptions of the nature of science: do they really influence teaching behaviour? *Science Education* **71** (5) 721–734

Matthews M 1989 A role for history and philosophy in science teaching *Interchange* **20** (2) 1–2

Mead M and Metrau R 1957 The image of the scientist amongst high-school students *The Sociology of Science Education* ed Barber B and Hirsch W (New York: Free Press of Glencoe)

Millar R 1989 Bending the evidence: the relationship between theory and experiment in science education *Images of Science* pp 38–61 ed Millar R (Lewes: Falmer Press)

Pumfrey S 1989 The concept of oxygen: using history of science in science teaching *Teaching the History of Science* ed Shortland M and Warwick A (Edinburgh: Basil Blackwell) pp 142–155

Rutherford F, Holton G and Watson F 1970 *The Project Physics Course* (New York: Holt, Rinehart and Winston)

Schutz A and Luckmann T 1973 *Structures of the Life-World* (New York: Heinemann)

Selley N 1989 The philosophy of school science *Interchange* **20** (2) 24–32

Sherratt W 1982 History of science in the science curriculum: an historical perspective *School Science Review* **64** 225–236

Solomon J 1984 Prompts, cues and discrimination: the utilization of two separate knowledge systems *European Journal of Science Education* **6** 277–284

Solomon J 1991 *Exploring the Nature of Science* (Glasgow: Blackie)

Tamir P 1989 History and philosophy of science and biological education in Israel *Interchange* **20** (2) 95–98

Wandersee J 1986 Can the history of science help science educators anticipate students' misconceptions? *Journal of Research in Science Teaching* **23** 581–597

Winchester I 1989 History, science and science teaching *Interchange* **20** (2) i–vi

Science should appear to our pupils as a growing fabric of knowledge in which one piece that they learn reacts with other pieces to build fuller knowledge. We must be careful to introduce any piece that we teach with an indication of our purpose, saying clearly how we are trying to build more science. And after we have taught a piece of science we should look back on it and talk with our pupils about the way in which that piece fits in with the rest and builds more....

Think of the pupil who learns a piece of physics thoroughly; trying his own experiments, watching demonstrations, discussing with the teacher, doing his own thinking. He makes this knowledge his own, and says: 'I understand this'. That is a proud possession, giving a sense of power, a sense of strong knowledge which can be of lasting value in his education.

Eric Rogers *Revised Nuffield Physics General Introduction* p 63

CHAPTER 4

CHILDREN LEARNING SCIENCE

Joan Bliss

4.1. BACKGROUND

The past thirty years have seen worldwide investment in changes in the teaching of science. The 1960s saw large scale curriculum development in the USA in the sciences, in physics (PSSC), chemistry (CBA and ChemStudy) and biology (BSCS), followed in England by more than a dozen projects sponsored by the Nuffield Foundation, of which Nuffield Physics directed by Eric Rogers was an early and influential one. Many other countries followed, adapting these ideas or developing their own.

Parallel developments in primary education in many countries were much influenced by the work of Piaget, amongst them a project at Berkeley directed by Robert Karplus (SCIS), projects in Britain in science and mathematics for ages 5–13, as well as the Australian Science Education Project (ASEP).

Despite all this work, and some say because of it, children still find learning science difficult. This chapter is about some of the reasons why they do. Understanding children's learning of science is important both for improving the teaching of science and, more generally, as a part of understanding learning and cognitive development.

During the 1960s and later, the work of Jean Piaget had a direct influence on thinking about children learning science. More recently new ideas from cognitive psychology, social psychology, language, anthropology and science education have begun to complement, modify or change our view of learning and teaching. Having had the privilege of working with Piaget for nearly a decade, and since I believe that there are important connections between his work and aspects of the newer approaches, I shall use his ideas as a point of reference for the discussion.

4.2.   PIAGET AND THE PRIMARY CURRICULUM

Piaget provided a rich and detailed collection of data about children's thinking and reasoning. He traced how their ideas about conservation of amount, volume, density, matter, number, length, perimeter and area develop. He described the development of their ideas about measurement, movement, distance, speed, time, space, geometry and chance. Those developing the primary school curriculum could use these ideas directly, and in consequence much of that curriculum (for example, children classifying objects) reflects Piaget's ideas, but with their origin often forgotten.

4.3.   PIAGET'S STAGE THEORY

Piaget was not content just to describe the many different ways in which children think and act as they develop. He wanted to know what their ideas about different things, such as the conservation of amount and of volume, had in common. Thus he hypothesized mental structures to account for changes in the way children think, and postulated a series of qualitatively different stages to describe children's intellectual development from birth to adolescence.

This gave educators a new way of thinking about teaching and learning. They used it as a tool to match the content of science curricula to children's spontaneous development, planning science activities to have a basis in what was known about development. In *Match and Mismatch* Harlen, Darwin and Murphy (1977a,b) attempted such a match of activities to development, so that their curriculum contains an ordering of concepts such as classification, amount, length, volume, the life cycle and cause and effect, and of processes such as observing, problem solving, raising questions, exploring and finding patterns in observations.

In secondary school science, Shayer in the 1970s used Piagetian stages of intellectual development to assess the Nuffield Foundation's projects in biology, chemistry and physics. He claimed:

First, the stages of the course ought to follow the same order of increasing logical complexity as are present in the pupils' own development. Secondly, the age range over which the course is taught should match the age range over which these stages develop.

(Shayer 1971 p 183)

Shayer distinguished two levels of matching:
• the minimum level the pupil must have attained for interest to be sustained;
• the level needed to gain what is expected by the project.

He was able to give reasons why, for example, the early use in chemistry of the mole concept as an integrating idea was bound to present difficulties, recommending delaying its introduction (Ingle and Shayer 1971). In the Teachers' Guides for the Nuffield physics course, Eric Rogers wrote a good deal of advice to teachers about how children learn and think. Shayer wrote of it:

> *It is clear both from the spirit and the detail of the Nuffield physics course, particularly in the first three years, that the fact of conceptual development in children was both known and allowed for in a more conscious way than in the planning of the chemistry or biology courses....*
>
> (Shayer 1972 p 31)

Such analyses were important in creating an awareness of the sources of difficulty of some science concepts. But they also created a tendency to say that one must not teach something until children are intellectually 'ready' for it, permitting opting out of teaching 'difficult' ideas. Bruner (1968) pointed out that 'readiness' is a mischievous half truth because

> *One teaches readiness or provides opportunities for its nurture; one does not simply wait for it.*
>
> (Bruner 1968 p 29)

Piaget never dealt with large numbers of subjects. His concern was philosophical, about how knowledge develops, illuminated by how children think. For him it was the fixed order of succession of cognitive stages that was crucial, not the age at which children reached a given stage, which he expected to depend on physical maturation, learning, social factors and motivation.

Shayer, Kucheman and Wylam (1976) and Shayer and Wylam (1978) carried out a large scale survey of secondary pupils between the ages of 11 and 16. By their criteria less than 30% of children in English comprehensive schools were in Piaget's formal operational stage at 16. Similar work in the USA (e.g. Lawson and Renner 1974, Rowell and Renner 1976) found many American undergraduates reasoning concretely rather than abstractly.

4.4.   FORMAL OPERATIONS AND SCIENTIFIC REASONING

Piaget saw the stage of formal reasoning as using general operational schemes, such as isolating and controlling variables, combinatorial, correlational, probabilistic and proportional thinking, and not in terms of reasoning specific to a domain.

This led to attempts both to investigate and to improve pupils' use of formal and scientific patterns of reasoning. Lawson and Wollman (1976) created reasonably effective teaching materials to promote 13-year-olds' understanding of the idea of a controlled experiment. Karplus, Lawson *et al* (1977) developed new science programmes designed to enhance the acquisition of scientific reasoning.

More recently, Shayer and Adey developed a two-year intervention programme with 11- and 12-year olds designed to accelerate formal thinking and promote general thinking skills (Adey, Shayer and Yates 1989; Shayer and Adey 1991, 1992). They get possible effects, in some cases, on public examination results some two or three years after the intervention without very clear positive effects on more immediate post-tests. Alternative explanations of these effects remain to be investigated.

Proportional reasoning, vital for science and mathematics, has attracted particular interest (for useful surveys see Karplus, Pulos and Stage 1981 and Tourniaire and Pulos 1985). Initially proportional reasoning was seen as a general cognitive structure. However it soon appeared that adolescents' proportional reasoning is very much tied to the nature of the task (Karplus and Peterson 1970, Karplus and Karplus 1972, Karplus, Karplus and Wollman 1974). Mellar and Bliss (1993) (also Mellar 1987) argue that proportional reasoning is more a collection of strategies than a unitary cognitive structure. Karplus, Karplus, Formisano and Paulsen (1975, 1977) found in a study in seven countries that differences between countries were much smaller than differences between social groups within countries. Kurtz and Karplus (1979) devised reasonably successful teaching schemes for proportional reasoning for 15–16 year olds.

4.5.   CONSTRUCTIVISM AND THE SCIENCE CURRICULUM

In the sense in which constructivism has been mainly understood in science education, Piaget was one of its early proponents, seeing children as constructing their own knowledge through their own activity. But while he followed this through in detail for younger children, accounts of older pupils' construction of ideas about matters of importance in science were lacking. Andersson diagnosed the difficulty as follows:

..although Piaget and others have described the development of reasoning in such a comprehensive manner, the model does not tell us, for instance, about what conceptions the pupils have of electricity, heat, light, matter, etc. before they begin science lessons. This is because Piaget and others have been interested in the development of reasoning in general. Our interest lies in specific subject areas. We must therefore to a large extent find out about the pupil's initial position by focusing on research activities on this problem.

(Andersson 1984)

It would be more correct to say that Piaget was interested in the knowledge which children could acquire without going to school, and his interest in the development of reasoning in general only came later. In fact Piaget does have a contribution to make to understanding pupils' ideas in areas relevant to secondary school science, in results from his exploration of ideas about cause and effect. Unfortunately this work was only published in a summary account in 1974 (though many detailed studies appeared in the *Études d'Epistemologie Génétique*).

From the 1970s a world-wide trend in science education developed, describing pupils' ideas in science. It became clear that their ideas were often very different from those of their teachers, or from the science specified in the curriculum. Duit (1989) took this trend to result from dissatisfaction with the results of curriculum development, noting that 70% of the research in Germany was in physics. Séré and Weil-Barais (1989) describe similar work in LIRESPT in Université Paris VII on electricity, light, temperature, heat, combustion, gaseous states, pressures, force and energy, again showing a bias towards the physical sciences. If science is difficult, physics appears especially so.

Such work often took a constructivist view of learning, in which learners are seen as taking an active role in the development of their own ideas. Children's idiosyncratic ideas were no longer seen as 'wrong', but merely as different from teachers' ideas (see Driver and Erickson 1983, Gilbert and Watts 1983 for overviews; and Driver, Guesne and Tiberghien 1985 for accounts of children's ideas).

The more that was found out about the surprising ways in which pupils construe the world, the more urgent became the desire to find ways of changing what they thought, especially in the face of the evidence that some of these intuitive ideas strongly resist change, even at university level (Viennot 1979). Attempts included mechanics (Champagne, Gunstone and Klopfer 1985), particle theory of matter (Nussbaum and Novick 1982) and general science (Hewson and Hewson 1983). The Children's Learning in Science project (1987) tried conceptual change strategies for energy, the particle theory of matter and plant nutrition. Overall, although strategies have varied widely, successes have broadly been modest.

Soon computers entered the fray. The Conceptual Change in Science project (Hennessy *et al* 1993) had limited success in getting young secondary students to reason in a Newtonian way about force and motion, using computer simulations of Newtonian and non-Newtonian 'realities', with pupils themselves able to give the rules governing motion. Some important pre-conceptions even increased. Newton, it seems, is difficult.

Ideas about how to effect change include cognitive conflict and the specification of conditions for conceptual change (Posner *et al* 1982). There are several things one can do with conflict: try something else, think about it, or block it. Bryant (1982) points out that to learn from contradictions requires more than simply being made aware of conflict, arguing the need to present alternatives (try something else). Others emphasize the need for reflection (think about it), for example Karmiloff-Smith (1984) argues that conceptual change comes from practical success, after which one tries to understand how that success was achieved. Conflict can be blocked by limiting conflicting ideas to distinct contexts.

Reflection means experimenting with one's own theorizing. In the Tools for Exploratory Learning programme, we used software tools to allow 11–14 year old pupils to construct models with which they could experiment with their own ideas (Bliss and Ogborn 1989, 1992a,b). Becoming aware of how well their ideas fitted a problem could enable them to see some of their limits.

4.6. CRITIQUE OF CONSTRUCTIVISM

Constructivism says that knowledge is constructed; beyond that it is a many-headed beast. For Piaget or for Kelly (1955) knowledge is constructed by the individual by individual means; for (say) Vygotsky (1978) it is constructed by the individual through social means; for sociologists such as Berger and Luckmann (1967) its construction and validation is social. Thus 'constructivism' gives different answers to the questions of who makes knowledge, how this is done, and on what basis it is held to be knowledge.

Although constructivism is a heterogeneous movement, many constructivists in science education derive their positions either from Piaget, or from Kelly. These two origins lead to quite distinct and different views of 'what knowledge is' and thus have very different implications for research and classroom practice (Bliss 1993).

In science education, Pope and Gilbert (1983) draw inspiration from Kelly's personal construct theory (by analogy, since Kelly was a therapist not an educator), using it to lend support

> ... to teachers who are concerned with the investigation of student views and who seek to incorporate these views within the teaching-learning dialogue and who see the importance of encouraging students to reflect upon and make known their construction of some aspects of reality'.

Their focus is on the individual and on the uniqueness of each person's construction of the world. Osborne and Wittrock (1985) spell out a crucial difficulty: this requires the teacher to diagnose each individual's constructs. In fact, however, most such research describes 'typical' constructs, not individual ones.

Such individual constructivism does not attribute a sufficient role to the teacher, the parent or the peer, and this has rightly led to attention being given to ideas from Vygotsky and others about the role of the adult or teacher in the pupil's learning.

Other constructivists draw inspiration from re-interpretations of Piaget, notably von Glasersfeld (1985, 1989). However, Piaget has nothing to say about individual differences, so although his philosophical ideas about the development of knowledge can be recruited, these leave open how one should think about the development of individuals. Teachers have to teach pupils, not 'epistemic subjects'. von Glasersfeld and Wheatley also seem to think that it follows from Piaget that the real world does not exist, being created by human thought and being dependent upon such thought.

From a constructivist perspective, knowledge originates in the learner's activity performed on objects. But objects do not lie around ready made in the world but are mental constructs.

(Wheatley 1991 p 10)

This is a total misinterpretation of Piaget's constructivism, which is plainly realist, with intelligence deriving from real actions on real objects.

These pages contain an account of an epistemology that is naturalist without being positivist; that draws attention to the activity of the subject without being idealist; that equally bases itself on the object, which it considers as a limit, therefore existing independently of us but never completely reached (known); and above all sees knowledge as a continuous construction.

(Piaget 1968)

In an excellent critical appraisal of constructivism, Matthews makes it clear that constructivism is often made into an ideology.

Constructivism in science and mathematics education is not just a theory of human concept formation and learning, nor just a theory of the way science develops and is validated, it is perhaps pre-eminently a theory of education (to use the title of one prominent constructivist's book). It is a view about how teaching should proceed, how children should be treated, how classrooms should be organized, how curricula should be developed and implemented, and sometimes even a view about the purposes of schooling.

(Matthews 1992)

A different objection to constructivism is that it sees the child just as learning about the world through experience. But science teaching can be seen as the way in which pupils are introduced into the communal world of science concepts and techniques, and communal standards of argument and evidence. Matthews gives an example: how Galileo changed not the facts but the arguments. Galileo did not see the real objects of science any differently from any one else. But he did describe them differently and it is these descriptions, the theoretical objects of science, with which pupils have to grapple. Constructivism rarely distinguishes between making personal sense of the real world, and understanding the socially constructed world of scientific ideas.

4.7. CHALLENGES TO PIAGET

Three challenges to Piaget are particularly relevant to science education.

They are:
- questioning the existence of 'formal operations';
- questioning the existence of general, non-domain specific abilities and of stages;
- insisting on the importance of the social, cultural and contextual dimensions.

Many doubts about formal thinking stem from work which shows how even highly educated adults perform badly on tasks involving abstract hypothetical thinking (Wason 1966, 1978, 1984; see Evans 1989 for a review). The tasks get much easier when they are translated into concrete contexts (for example, Johnson-Laird, Legrenzi and Sonino-Legrenzi 1972). Piaget (1977) conceded that people would differ in their formal reasoning according to aptitude and personal specialization. But this defence denies the essential context-free nature of formal reasoning.

A kind of thinking operating on symbols and signs, rather than on representations of objects and events, certainly exists. The question is whether it should be regarded as a special way of thinking used on some occasions for some purposes, rather than being universalized as a highly developed form of reasoning. There is little reason to suppose that it is much used in everyday life. Piaget may have made formal reasoning into a false God.

More generally, Keil (1986) and Carey (1985, 1990) see adults as different from children mainly by knowing more, not by possessing general cognitive structures. Carey argues that much of the evidence given to support stage-like development of cognitive structures could in fact reflect domain-specific reorganizations of knowledge. She holds however that there are some 'foundational concepts', for example causality, but she sees them not as providing general mechanisms for thinking but as merely being used differently at different points in time.

Others, such as Cheng and Holyoak (1985, 1989), see children (and adults) as using pragmatic rather than inferential reasoning for typical everyday tasks. Girotto and Light (1992) discuss context-sensitive rules, limited in their use by pragmatic forms of permission and obligation, expertise in which is acquired early in life (Turiel 1983).

Finally, there is the socio-cultural challenge which sees context and cultural practice as the fundamental units within which thought has to be analysed. Human mental functioning is seen as emerging from and located in social practices. Such a theory of culture and cognition resists the separation of the individual from the environment of daily life (Laboratory of Comparative Human Cognition 1983). Children's apparent abilities to solve problems can depend on whether they encounter them in school or in the context of everyday practical activities (Lave, Murtaugh and Rocha 1988; Newman, Griffin, Cole 1989). A consequence of this new focus is to cast doubt on the very notion of the transfer of knowledge from one context to another. 'Transfer' now involves crossing contextual barriers (Gick and Holyoak 1980). In these theories development reduces to learning the culture.

4.8.   MENTAL MODELS

In the last decade much attention has been given to human thinking and reasoning seen as the manipulation of imagined entities, as using mental models. This view stands against 'formal reasoning', that is, reasoning seen as logical processing through the use of inference rules, but accords better with what Piaget described as concrete operational reasoning. There is a good case for saying that Piaget undervalued too much his account of concrete reasoning, seeing it only as a step towards formal reasoning.

One kind of work under the label 'mental models' is that of Johnson-Laird (1983) and Johnson-Laird and Byrne (1991). For them a mental model is more like a set of mental tokens which are manipulated to form a structure which might correspond with a structure in reality. Their mental models do not directly represent reality. They use this idea to attack the notion of 'mental logic' or of a 'language of thought'.

Another kind of work on mental models derives from attempts in artificial intelligence to construct systems which reason 'naturally'. An example is the work of de Kleer and Brown (1983) on a system designed to work out what a device such as a pump might do, given a description of its parts. Their mental model is built up in four steps:

- the system builds a description of the structure of the device;
- the system works out a number of ways in which the device might function, building a causal model of it, in a process called *envisioning*;
- the system compares these possibilities with what happens, and may revise its model as a result.

Gutierrez and Ogborn (1992), used these ideas to develop a causal framework for analysing some of the data from pupils' alternative conceptions in mechanics. One of the advantages of their analysis is that it accounts not just for fixed ways of thinking but also for changes of mind, obviously important for learning.

Piaget never uses the term 'mental model', but can often be seen as speaking a similar language. Thus leaving the sensory-motor stage requires the construction of a new plane of reality, of representation in place of direct action. Similarly concrete operational thinking operates on imagined entities; that is, young children think about the world in terms of imagined objects and events. What Piaget rarely does is to discuss the nature of these representations.

4.9. FOCUS ON ONTOLOGY

Much past work on learning has drawn its ideas from epistemology, that is, the study of the grounds of knowledge. For example 'direct experience' has been seen as basic both to learning and to the foundations of knowledge. The mental models position points in a new direction, namely an ontological focus on studying what people take to be the nature of the things in the world around them. What *is* comes before what *is the case*.

The work of Keil (1979, 1981) and of Carey (1985) is in this mould. They suggest that the conceptual system is articulated by a core of ontologically basic categories, such as *physical object* and *event*. Basic ontological categories are few, but provide the core organizational system for other concepts. Carey suggests that domain specific knowledge emerges from people's naive theories. An example is her account of how children's biological ideas emerge from their naive psychology.

I shall now discuss some work, much of it done in a group in London with which I have been closely associated. It attempts to bring together:

• a description of common sense everyday reasoning,
• a description of basic ontological categories.

One series of studies (Mariani and Ogborn 1990, 1991, Mariani 1992, Ogborn 1992) elicits underlying common sense categories of pupils (aged 8 to 18 plus) which serve to organize thinking about such things as the Sun, oxygen, matter, stars, space, water, movement. There seems to be a fundamental 'ontological space' with a small number of stable ontological dimensions along which objects and events can be placed: dynamic versus static, place-like versus localized, cause versus effect, discrete versus continuous. Further work looks at the ontology of events, relating them to actions.

An extended series of studies has tried to get at ideas underlying people's thinking about force and motion. Ogborn (1985) proposed an account deriving from Hayes (1979, 1985) in terms of effort and support. This led to the identification of a number of prototypical motions, such as *put, fall, push/pull, roll/slide, carry, throw, walk/run, jump, fly*, and to showing that people consistently use such prototypes to recognize and account for motions (Bliss, Ogborn, Whitelock 1989, Whitelock 1990, Law 1990, Whitelock 1991). Now we have proposals for how such prototypes might develop in the tacit thinking of children from a very early age (Bliss and Ogborn 1990, 1992c, 1993).

The work of the London Group takes ontology to be fundamental. Ogborn (1992) glosses common sense reasoning as, 'Ordinary, everyday, unreflective, practical reasoning about things which are mostly seen as obvious'. It relies on ontological judgments of the world, about the way we imagine the basic nature of objects and events, which tells us what to expect them obviously to be able to do. 'Obviousness' replaces 'logical necessity'. This thinking also draws on Piaget, in whose last books (Piaget and Garcia 1983, 1987), meanings were tied unambiguously to the nature of things.

Two meanings of an object are, subjectively, what can be done with it and, objectively, what it is made of or how it is composed.

(Piaget 1983 p 58)

4.10. A WAY FORWARD?

In this section I will sketch my own view of how we might get further in understanding mental models.

Piaget saw knowledge as derived from two kinds of activity, logico-mathematical and physical activity, but gave only minimal attention to the latter. Through logico-mathematical activity actions turn inwards to become mental actions. This activity coordinates the child's actions on objects rather than focusing on the objects themselves. He calls this process of reflecting on actions 'reflective abstraction'. He emphasizes that mental operations derive from the most general types of actions such as ordering, uniting, classifying. Such actions, by becoming mental operations, form the basis of the child's cognitive structure.

Physical activity, on the other hand, is about acting on objects and physical situations, and learning about their size, colour, shape, weight, texture, smell etc. Piaget calls this process 'empirical abstraction'. Being less interested in empirical abstraction, he tended to treat it when he met it as a source of difficulty rather than as a new line of thought to pursue. Thus in the introduction to his study of causality, Piaget (1974) states that this area provided many more difficulties than anticipated, particularly the importance of specific differences between particular physical realities,

*Each new analysis threatened to contradict as well as complete some of the preceding ones since, let us repeat, the causal explanations **depend more on the objects than on the subjects.***

(Piaget 1974 p 3, my emphasis)

He constantly examines logical schemes applied to the physical world to organize and understand it. Such a way of viewing thinking is to neglect some of what is special about understanding the physical world, especially aspects important in science.

So it seems to me that we should pay much more attention to empirical abstraction, to physical rather than logical schemes. Physical schemes would not be context-free. Indeed they would have to do with the creation of contexts. Contrary to some received wisdom related to the desirability of transfer, it is cognitively rather sensible to be very cautious about generalizing across different contexts. So we might expect a slow, rather conservative process of noticing salient features of new contexts, which would generate new and different constraints. Gradually such schemes would abstract some of the features common to many physical situations, but they would also contain the constraints of particular situations.

Examples might be schemes concerning rigidity, resistance, or flow. A scheme for rigid objects must allow that, although many objects can be understood readily in this way, there are some which fit it partially (hard india-rubbers or breakable tea-cups) and some which fit it well only sometimes (ice-cubes but only when cold).

More general physical schemes might be to do with containment, substitution and joining. These are important because of their importance in reasoning with mental models. Reasoning about sets can largely be done in terms of containers, for example, substituting one imagined entity for another, or joining entities into compound entities, each have evident uses in many sorts of reasoning. At the same time such reasoning is, and should be, constrained by whether the real entities being thought about can, of their nature, be manipulated in this way in a given context.

4.11.   CONCLUSIONS

Piaget took epistemology seriously. For him the psycho-genesis of knowledge is necessary to epistemology, asking how it comes about that people generally come to see the world as they do. He showed us that children brought to school a whole wealth of personal knowledge and ways of thinking about the world.

It is also crucial to understand the processes by which children's ideas about the world may change or resist change. We should focus more than we have on the role of the teacher, rather than seeing learning only as the individual child making sense of experience. We need to understand much better the nature and importance of context, which might clarify our ideas about why many of pupils' naive theories are resistant to teaching, about the difficulties of transfer of learning, and about the importance or not of general developmental mechanisms. All of these problems are particularly important to science.

To understand something is more than seeing why it might be true (epistemology). The mental models position, which tries to understand what people take to be the nature of the things in the world around them (ontology), has I think much to offer for a better understanding of the problems of learning science. For instance, the Newtonian view of force and motion involves a complete imaginative reconstruction of causes of motion (Bliss, Morrison and Ogborn 1988). Science teaching consists in part of educating the intuitions.

We need to know where other areas of science involve similar reconstructions. More work needs to go into an analysis of the origins of children's informal ideas across other science topics, to see how far they relate to deep-rooted intuitive thinking, and to see to what extent the scientific account itself challenges that thinking. We need to understand the gap, when there is one, between the everyday common sense ontology which children are using and the scientific ontology.

Learning science is not about making sense of the real world purely for oneself; it is about making sense of how scientists have made sense of the world. The pupil's task in school is to come to terms with the scientific account of the world. But while scientists can see how the different ideas inter-relate and fit into more general theories, pupils meet these different ideas in isolated, separate contexts which may or may not relate to their own experiences.

Teaching science is a matter of conveying a mental model. And teaching *about* science is a matter of conveying that science consists of stories which have been made up in the hope of describing essential bits of the nature of things. Both need the exercise of imagination through analogy and metaphor, together with plenty of first hand knowledge of phenomena so that physical schemes have a chance to get built up and be applied to new things. At all of these—imaginative talk and providing rich experiences of reality—Eric Rogers was a past master.

REFERENCES

Adey P, Shayer M and Yates C 1989 *Thinking Science: the Curriculum Materials of the CASE Project* (Basingstoke: Macmillan Education)

Andersson B 1984 A framework for discussing approaches and methods in science education *Educational Research Workshop on Science in Primary Education* Council for Cultural Cooperation, Edinburgh p 984

Berger P and Luckmann T 1967 *The Social Construction of Reality* (London: Penguin)

Bliss J Morrison I and Ogborn J 1988 A longitudinal study of dynamics concepts *International Journal of Science Education* **10** (1) pp 99–110

Bliss J, Ogborn J and Whitelock D 1989 Secondary school pupils' common sense theories about motion *International Journal of Science Education* **11** (3) pp 261–272

Bliss J and Ogborn J 1989 Tools for exploratory learning *Journal of Computer Assisted Learning* **5** pp 37–50

Bliss J and Ogborn J 1990 A psycho-logic of motion *European Journal of psychology of education* **5** (4) pp 379–390

Bliss J and Ogborn J 1992a Reasoning supported by computational tools *Computers in Education* **18** (1–3) pp 1–9, reprinted in *Computer Assisted Learning* ed Kibby M R and Hartley J R (New York: Pergamon Press) p 1–9

Bliss J and Ogborn J 1992b *Tools for Exploratory Learning End of Award Reports to ESRC*:
Executive Report
Summary Report
Technical Report 2: Semi-quantitative Expressive Modelling
Technical Report 3: Semi-quantitative Exploratory Modelling
Technical Report 5: Qualitative Modelling
Technical Report 6: Tasks

Bliss J and Ogborn J 1992c Steps towards a formalization of a psycho-logic of motion *Intelligent Learning Environments and Knowledge Acquisition in Physics* ed Tiberghien A and Mandl H NATO ASI Series (Berlin: Springer Verlag) pp 65–89

Bliss J and Ogborn J 1993 Steps towards a formalization of a psychologic of motion *Journal of Intelligent Systems* **3**

Bliss J 1993 The relevance of Piaget to research into children's conceptions *Childrens' Informal Ideas About Science* ed Black P and Lucas A (London: Routledge) pp 20–44

Bruner J 1968 *Towards a Theory of Instruction* (New York: Norton)

Bryant P 1982 'The role of conflict and agreement between intellectual strategies in children's ideas about measurement' *British Journal of Psychology* **73** pp 243–251

Carey S 1990 Are children fundamentally different kinds of thinkers and learners than adults? *Cognitive Development to Adolescence* ed Richardson K, Sheldon S and Hove E (Sussex: Lawrence Erlbaum)

Carey S 1985 *Conceptual Change in Childhood* (Cambridge, Mass: MIT Press)

Champagne A, Gunstone R and Klopfer L 1985 Effecting changes in cognitive structures among physics students *Cognitive Structure and Conceptual Change* ed West L and Pines L (New York: Academic Press)

Cheng P and Holyoak K 1985 Pragmatic reasoning schemas *Cognitive Psychology* **17** pp 391–416

Cheng P and Holyoak K 1989 On the natural selection of reasoning theories *Cognition* **33** pp 285–313

Children's Learning in Science (CLISP) 1987 *CLIS in the Classroom: Approaches to Teaching* Centre for Studies in Science and Mathematics Education, University of Leeds

de Kleer J and Brown, J 1983 Assumptions and ambiguities in mechanistic mental models, in Gentner D and Stevens A *Mental Models* (New Jersey: Lawrence Erlbaum)

Evans J 1989 *Bias in Human Reasoning: Causes and consequences* (New Jersey: Laurence Erlbaum)

Driver R and Erickson G 1983 Theories-in-action: some theoretical and empirical issues in the study of students' conceptual frameworks in science *Studies in Science Education* **10** pp 37–60

Driver R, Guesne E and Tiberghien A (ed) 1985 *Children's Ideas in Science* (Milton Keynes: Open University Press)

Duit R 1989 Research on students' conceptions in science—perspectives from the Federal Republic of Germany *Adolescent Development and School Science* ed Adey P, Bliss J, Head J and Shayer M (Lewes: The Falmer Press)

Gilbert J and Watts M 1983 Concepts, misconceptions and alternative conceptions: Changing perspectives in science education *Studies in Science Education* **10** pp 61–98

Girotto V and Light P 1992 The pragmatic bases of children's reasoning *Context and Cognition: Ways of Learning and Knowing* ed Light P and Butterworth G (London: Harvester Wheatsheaf)

Glasersfeld E von 1989 Cognition, Construction of Knowledge and Teaching *Synthese* **80** pp 121–140

Glasersfeld E von 1985 An interpretation of Piaget's constructivism *Revue Internationale de Philosophie* **36** pp 612–635

Gick M and Holyoak K 1980 Analogical problem solving *Cognitive Science* **12** pp 306–355

Gutierrez R and Ogborn J 1992 A causal framework for analysing alternative conceptions *International Journal of Science Education* **14** (2) pp 201–220

Harlen W, Darwin A and Murphy P 1977a *Match and Mismatch. Raising Questions* (Edinburgh: Oliver and Boyd for the Schools Council)

Harlen W, Darwin A and Murphy P 1977b *Match and Mismatch. Finding Answers* (Edinburgh: Oliver and Boyd for the Schools Council)

Hayes P 1979 The naive physics manifesto *Expert Systems in the Micro-Electronic Age* ed Michie D (Edinburgh: University of Edinburgh Press)

Hayes P 1985 The second naive physics manifesto, in *Formal Theories of the Common Sense World* ed Hobbs J and Moore R (New Jersey: Ablex)

Hennessy S *et al* 1993 A classroom intervention using a computer-augmented curriculum for mechanics *International Journal of Science Education* (in press)

Hewson M and Hewson P 1983 The effect of instruction using students' prior knowledge and conceptual change strategies on science learning *Journal of Research in Science Teaching* **20** (2) pp 731–743

Ingle R and Shayer M 1971 Conceptual demands in Nuffield O-level chemistry *Education in Chemistry* **8** pp 182–183

Johnson-Laird P, Legrenzi P and Sonino-Legrenzi M 1972 Reasoning and sense of reality *British Journal of Psychology* **63** pp 395–400

Johnson-Laird P 1983 *Mental Models* (Cambridge: Cambridge University Press)

Johnson-Laird P and Byrne M 1991 *Deduction* (New Jersey: Lawrence Erlbaum)

Karmiloff-Smith A 1984 Children's problem solving *Advances in Developmental Psychology* ed Lamb M, Brown A and Rogoff B (New Jersey: Lawrence Erlbaum) pp 39–90

Karplus R and Peterson R 1970 Intellectual development beyond elementary school II: ratio, a survey *School Science and Mathematics* **70** pp 813–820

Karplus, R and Karplus E 1972 Intellectual development beyond elementary school III: ratio, a longitudinal study *School Science and Mathematics* **72** pp 735–742

Karplus E, Karplus R and Wollman W 1974 Intellectual development beyond elementary school IV: ratio, the influence of cognitive style *School Science and Mathematics* **74** pp 476–482

Karplus R, Karplus E, Formisano M and Paulsen A 1975 Proportional reasoning and control of variables in seven countries *Advancing Education through Science Programs* Report ID-65

Karplus R, Karplus E, Formisano M and Paulsen A 1977 A survey of proportional reasoning and control of variables in seven countries *Journal of Research in Science Teaching* **14** pp 411–417

Karplus, R Lawson A E *et al* 1977 Science teaching and the development of reasoning. Lawrence Hall of Science, University of California, Berkeley, California (Workshop materials in five parts: physics, chemistry, biology, earth science and general science)

Karplus R, Pulos S and Stage E 1981 Early adolescents' structure of proportional reasoning *Proceedings of Fourth International Conference for the Psychology of Mathematics Education* Berkeley, California pp 136–142

Keil F 1986 On the structure-dependent nature of stages of cognitive development *Stage and Structure—Reopening the Debate* ed Levin I (Norwood, NJ: Ablex)

Keil F 1979 *Semantic and Conceptual Development: An Ontological Perspective* (Cambridge, Mass.: Harvard University Press)

Keil F 1981 Constraints on knowledge and cognitive development *Psychological Review* **88** (3) pp 197–227

Kelly G 1955 *The Psychology of Personal Constructs* **1** and **2** (New York: Norton)

Kurtz B and Karplus R 1979 Intellectual development beyond elementary school VII. Teaching for proportional reasoning *School Science and Mathematics* **79** pp 387–398

Laboratory of Comparative Human Cognition 1983 Culture and cognitive development *Mussen's Handbook of Child Psychology* ed Kessen W 4th Edition **1** (New York: Wiley)

Lave J, Murtaugh M and de la Rocha O 1984 The dialectic of arithmetic in grocery shopping *Everyday Cognition: Its Development in Social Context* ed Rogoff B and Lave J (Cambridge, Mass.: Harvard University Press)

Law N 1990 Eliciting and understanding common sense reasoning about motion *PhD Thesis* University of London

Lawson A and Wollman W 1976 Encouraging the transition from concrete to formal cognitive functioning: an experiment *Journal of Research in Science Teaching* **13** pp 413–430

Lawson A and Renner J 1974 A quantitative analysis of responses to Piagetian tasks and its implications for curriculum *Science Education* **58** pp 544–559

Light P and Butterworth G ed 1992 *Context and Cognition: Ways of Learning and Knowing* (London: Harvester Wheatsheaf)

Mariani M-C, Ogborn J 1990 Common sense reasoning about conservation: the role of events *International Journal of Science Education* **12** (1) pp 55–66

Mariani M-C, Ogborn J 1991 Towards an ontology of common sense reasoning *International Journal of Science Education* **13** (1) pp 69–85

Mariani, M-C 1992 Some dimensions of common sense reasoning about the physical world *PhD Thesis* University of London

Matthews M 1992 Constructivism and empiricism: an incomplete divorce (paper personally communicated)

Mellar H 1987 The understanding of proportion in young adults: investigating and teaching a formal level skill through a Logo microworld *PhD Thesis* University of London

Mellar H and Bliss J 1993 Expressing the student's concepts versus exploring the teacher's: issues in the design of microworlds for teaching *Journal of Educational Computing Research* **9** (1) pp 89–113

Newman D, Griffin P and Cole M 1989 *The Construction Zone—Working for Cognitive Change in School* (Cambridge: Cambridge University Press)

Nussbaum J and Novick S 1982 Alternative frameworks, conceptual conflict and accommodation: towards a principled teaching strategy *Instructional Science* **11** pp 183–200

Ogborn J 1985 Understanding students' understandings: an example from dynamics *European Journal of Science Education* **7** (2) pp 141–150

Ogborn J 1992 'Fundamental dimensions of thought about reality: object, action, cause, movement, space and time' in *Teaching About Reference Frames: From Copernicus to Einstein* Proceedings of GIREP Conference, Torun, Poland, August 1991 (Torun: Nicholas Copernicus University Press)

Osborne R and Wittrock M 1985 The generative learning model and its implications for science education *Studies in Science Education* **12** pp 59–87

Piaget J 1968 *Le Structuralisme* (Paris: Presses Universitaires de France)

Piaget J 1974 *Understanding Causality* (New York: Norton)

Piaget J 1977 Intellectual evolution from adolescence to adulthood *Thinking* ed Wason P and Johnson-Laird P (Cambridge: Cambridge University Press)

Piaget J and Garcia R 1983 *Psychogenese et Histoire des Sciences* (Paris: Flammarion)

Piaget J and Garcia R 1987 *Vers une Logique des Significations* (Geneva: Murionde)

Pope M and Gilbert J 1983 Personal experience and the construction of knowledge in science *Science Education* **67** pp 193–203

Posner G, Strike K, Hewson P and Gertzog W 1982 Accommodation of a scientific conception: Toward a theory of conceptual change *Science Education* **66** (2) pp 211–227

Rowell J and Renner J 1976 Quantity conceptions in university students: another look *British Journal of Psychology* **6** pp 1–10

Séré M and Weil-Barais A 1989 Physics education and students' development *Adolescent*

Development and School Science ed Adey P, Bliss J, Head, J and Shayer M (Lewes: The Falmer Press)

Shayer M 1971 How to assess science courses *Education in Chemistry* **7** pp 182–186

Shayer M 1972 Conceptual demands in the Nuffield O-level physics course *School Science Review* **54** (186) pp 26–34

Shayer M, Kucheman D and Wylam D 1976 The distribution of Piagetian stages of thinking in British middle and secondary school children *British Journal of Educational Psychology* **46** pp 164–173

Shayer M and Wylam D 1978 The distribution of Piagetian stages of thinking in British middle and secondary school children. 11–14 and 14–16 year olds and sex differentials *British Journal of Educational Psychology* **48** pp 62–70

Shayer M and Adey P 1991 Accelerating the development of formal thinking in middle and high school students II: post-project effects on science achievement *Journal of Research in Science Teaching*

Shayer M and Adey P 1992 Accelerating the development of formal thinking in high school students III: post-project effects on public examinations in science and mathematics *Journal of Research in Science Teaching*

Tourniaire F and Pulos S 1985 Proportional reasoning: a review of the literature *Educational Studies in Mathematics* **16** pp 181–204

Turiel E 1983 *The Development of Social Knowledge* (Cambridge: Cambridge University Press)

Viennot L 1979 Spontaneous reasoning in elementary dynamics *European Journal of Science Education* **1** (2) pp 205–221

Vygotsky L S 1978 *Mind in Society: The Development of Higher Psychological Processes* (Cambridge, Mass.: Harvard University Press)

Wason P 1966 *Reasoning: New Horizons in Psychology* ed Foss B (Harmondsworth: Penguin)

Wason P 1978 Hypothesis testing and reasoning, Unit 25, Block 4 *Cognitive Psychology* (Milton Keynes: Open University Press)

Wason P and Green D 1984 Reasoning and mental representation *Quarterly Journal of Experimental Psychology* **36**A pp 597–610

Wheatley G 1991 Constructivist perspectives on science and mathematics learning *Science Education* **75** (1) pp 9–22

Whitelock D 1990 Common sense understandings of causes of motion *PhD Thesis* University of London

Whitelock D 1991 Investigating a common sense model of causes of motion with 7 to 16 year old pupils *International Journal of Science Education* **13** (3) pp 321–340

If we are teaching for a sense of understanding science, we should ask questions that inquire visibly into the students' knowledge, ask for reasoning, ask for clear understanding, ask them to describe scientific work. In short, we should give them problems that they can answer if—but only if—they are following the course and achieving some of our aims. Obviously that 'if—but only if' is an ideal of examining that we can only strive towards.

Further, since our most important aims are long-term ones, for benefits that may not mature for months or years, our examinations cannot be tests of full success. The best we can do is to make them encourage success rather than prevent it.

Eric Rogers *Teaching Physics for the Inquiring Mind* p 68

CHAPTER 5

HUMANE AND HELPFUL ASSESSMENT

Paul Black

5.1. INTRODUCTION

The effect of the examination on the primary schools is by common consent disastrous. In the words of a primary head teacher who had recently returned from teaching in another country, 'It hinders true development and deprives the children of both understanding and enjoyment.' Or, as a secondary head put it, 'Five years of cramming stifles the eagerness to find out . . . when boys come here they are no longer interested in work '.

In order to cover the curriculum teachers press ahead whether or not the children have acquired a skill or understood a concept. 'We have to drive on even if the children haven't grasped what is being taught', lamented another primary head.

The effects on the secondary schools are no less deleterious.

(Dockrell 1991)

These words were written by a visitor who had studied the examination system in one developing country. Contrast them with a description of two teachers chosen because of their reputation for teaching for understanding:

... the key to the successes of Doug and Gary was associated with the way they were able to monitor for understanding.

(Tobin and Garnett 1988)

These two extracts reflect themes which were central to Eric Rogers' thinking about examinations, seen in the Nuffield O-level Physics Guides and in his Oersted Lecture, reprinted in this volume (Rogers 1969). The title

63

of the latter—Examinations: powerful agents for good or ill in teaching—makes his position clear. He spells out the need to reflect our values and priorities in our testing, and also the need to use testing to help students in learning with understanding. In this chapter, I shall try to pursue and develop both of these themes with reference to current developments in thinking about assessment and testing.

5.2.   THE INHERITANCE—NUFFIELD PHYSICS EXAMINATIONS

Eric Rogers set guidelines and standards for examinations. He was a practitioner—his writing was liberally illustrated with examples and his workshops for training examiners were renowned (see chapter 17 in this volume). Whilst I did not work on the O-level examinations, and my own work with the A-level physics examination was not under his direct influence, his indirect influence on us was marked.

The Nuffield A-level physics examination (Dobson 1985) is deliberately designed with varied components:

- a multiple-choice question paper;
- problem-solving questions, each needing a short but thoughtful response;
- questions based on reading about a new topic, requiring understanding of physics applied to making sense of the text;
- a data analysis exercise;
- discussion of the physics underlying a number of ideas or phenomena presented as short paragraphs;
- practical problems, that is, 'questions with apparatus', requiring thinking about observations, and manipulating simple apparatus;
- a practical investigation in which candidates carry out in school time an extended investigation into a topic of their own choice;
- a research and analysis exercise in which candidates, in school time, read and collect information about a topic or issue of their own choice.

The last two, each based on course-work for about two weeks of physics time, are assessed by the teacher, using criteria supplied by the Examination Board, samples being checked by the Board. Both require students to think, plan, test, check, reflect and review. They choose their own resources and areas of interest and can make mistakes, change their minds, try again. This takes time, and cannot be done with a timed test in an examination hall. The aim is to encourage original thinking, and the ability to make and carry out plans.

This collection of instruments was planned with three principles in mind, all of which reflect priorities in Eric Rogers' thinking. The first was to reflect and reinforce the aims of the course—which stressed making students active and thoughtful learners. The second was a principle of variety, so that difficulties some students may have with one kind of work can be compensated by evidence of strength in another. The third was to involve teachers in sharing responsibility for examining, using their direct knowledge of pupils' classroom work.

Recently, an international group met to prepare a review of examinations in eleven countries (Black 1993). Representatives from other countries were surprised at the Nuffield examination—and envious of it. Their testing was dominated by standard questions: a recall part followed by a mathematical problem, which reflect a very narrow view of ways to test understanding. Eric Rogers frequently pointed out how students can tackle such 'problems' with little understanding, and showed how they might be made more effective— mainly by asking students what the calculations meant.

5.3. UNDERSTANDING COMES FIRST—AT LAST!

Eric Rogers was very critical of the use of multiple choice questions. But he allowed that they could play a limited role and he was expert in training teachers to use them well and to avoid their many pitfalls. His experience of the dominance of this style of testing in the USA must have been a powerful influence. His message, that they could suppress learning with understanding, was prophetic.

Technical expertise in the use of multiple choice questions in standardized tests has reached a higher level in the USA than anywhere else in the world. However, many States are now abandoning them, because it is evident that they have done almost nothing to improve education. Since 1989, 16 States have started to develop alternative forms of assessment in science (20 in mathematics), and further new initiatives are developing rapidly (Blank and Dalkilic 1992). The new interest is in 'performance tests' which are closer to good classroom practice, in which teachers can be fully involved.

These changes are accompanied by fundamental criticisms of common assumptions about testing (Resnick and Resnick 1992). One concerns decomposability—it was assumed that a complex skill can be taught by breaking it into pieces and teaching and testing these separately. A second is decontextualization—it was assumed that something common to many contexts can be taught most economically by presenting it in abstract isolation. Both assumptions lead to testing with short 'atomized' questions, yet neither appears to be valid for most learners.

A further assumption is that of fixed general intelligence, expecting a single IQ test to indicate what an individual may hope to achieve. In the light of current evidence about the complex nature of intelligence (Sternberg 1992) this seems dangerously naïve. Eric Rogers gave his own warning— that IQ was only a measure of ability with an IQ test, which might correlate with performance in other tests only because of the poor quality of such tests. The emphasis now must be on using tests to identify and understand individual needs rather than to predict.

A collection of studies for the USA National Commission on Testing and Public Policy (Gifford and O'Connor 1992) illustrates the concerns which underlie this change of direction in testing. Professor Lorrie Shepard's closing summary states the essential point.

> *...all learning involves thinking. It is incorrect to believe, according to old learning theory, that the basics can be taught by rote followed by thinking and reasoning... even comprehension of simple texts requires a process of inferring and thinking about what the text means. Children who are drilled in number facts, algorithms, decoding skills or vocabulary lists, without developing a basic conceptual model or seeing the meaning of what they are doing, have a very difficult time retaining information (because all the bits are disconnected) and are unable to apply what they have memorized (because it makes no sense).*

(Shepard 1992)

The collection contains a critique of the multiple choice or very short answer tests which were until recently almost the only form of testing in USA schools.

> *Children who practice reading mainly in the form in which it appears in the tests—and there is good evidence that this is what happens in many classrooms—would have little exposure to the demands and reasoning possibilities of the thinking curriculum.*
>
> *Students who practised mathematics in the form found in the standardized tests would never be exposed to the kind of mathematical thinking sought by all who are concerned with reforming mathematical education.*

(Resnick and Resnick 1992)

The same article goes on to stress the key consequence to be drawn from the inevitability of teaching to the test.

> *Assessments must be so designed that when you do the natural thing— that is, prepare the students to perform well—they will exercise the kinds of abilities and develop the kinds of skills that are the real goals of educational reform.*

It commends examples of forms of assessment (many of them British) which

> *...could not only remove current pressures for teaching isolated collections of facts and skills but also provide a positive stimulus for introducing more extended thinking and reasoning activities in the curriculum.*

One might be forgiven for concluding that it has taken about twenty years for academic theories of testing to catch up with the good practice that Eric Rogers helped so powerfully to establish.

5.4. TEACHERS' FORMATIVE ASSESSMENT IN PRACTICE

In his Oersted lecture (reprinted, this volume), Eric Rogers spoke of 'mirror tests' to show students what they had learnt, to provide students with landmarks, and to show teachers what students had learnt. It was characteristic that he judged a liberal supply of questions more essential to Nuffield O-level than a textbook. 'Mirror tests' show Eric Rogers' concern with the formative role of assessment.

High-stakes testing tends to dominate classroom work and can so distort teaching as to destroy the possibility of any type of formative assessment. The detrimental effects of narrow external testing on science teaching have been amply recorded in a range of research studies.

- science is reduced to isolated facts and skills;
- the cognitive level of classroom work is lowered;
- pupils have to 'cover ground' at too great a pace for effective learning;
- teaching time is devoted to test preparation; inhibiting students' questions;
- learning follows testing, focusing on aspects that are easy to test;
- laboratory work stops unless there are laboratory tests;
- creative, innovative methods and topical content are dropped;
- teachers' autonomy is constrained, their methods revert to a uniform style and they are led to violate their own standards of good teaching (Duschl and Wright 1989, Herr 1992, Smith *et al* 1992, Tobin *et al* 1988b, Yager and McCormack 1989, Wood 1991).

It is not surprising that teachers dislike and suspect assessment. Few teachers have positive attitudes towards assessment, and good examples of formative assessment used in a programme of effective learning are hard to find.

Hodson (1986) describes school science departments claiming to place a high value on pupils' reactions to courses, yet appearing to have no means to assess these. Only 18% used tests for judging the effectiveness of teaching and learning, most using tests to stream pupils and to identify those with special learning needs. Assessment results ranked lowest of all in factors influencing curriculum decisions. Fewer than 12% of department heads claimed that they often identified the particular skills they were assessing.

Many studies show teachers thinking mainly in terms of summative assessment alone. Scott (1991) found many secondary teachers formalizing course-work assessment and, in so doing, breaking any possible connection with learning development. A similar story is told by Harlen and Qualter (1991) for primary school teachers responding to the new assessment demands of the National Curriculum in England. The overwhelming tendency was to separate assessment from information to be used in teaching.

5.5. TEACHERS' FORMATIVE ASSESSMENT—DOES IT MATTER?

Well-known studies (e.g. Mortimore *et al* 1988) emphasize that feedback, communication and record keeping are key aspects of the effectiveness of a school. Performance on external tests does improve with greater use of classroom testing (see Bangert-Drowns *et al* 1991, Rudman 1987), but the effectiveness of the feedback depends on its quality, in particular on whether it gives evidence of the nature of a student's difficulty rather than just signalling a failure.

In more radical changes feedback assessment is given a central learning role. The outstanding example is mastery learning. The salient points to emerge (Block *et al* 1989) are:

- mastery gives substantial learning gains;
- the key to achieving even larger gains is to enhance students' own active role in monitoring their own progress;
- effective mastery learning goes with clear aims, operationalized in learning activities and in relevant tests, breakdown of material into suitable learning units, formative tests linked to clear standards, with corrective materials and matching summative tests at the end;
- day-to-day formative assessment is a different art from summative assessment; both are needed but the scores of pupils on the two types may not be related;
- the variability in learning in a group may be reduced by helping the initially disadvantaged but there is controversy over whether such students then learn at the same rate as the rest or whether this only happens if the more able are held back.

Substantial benefits were obtained in an adaptation of mastery learning in Scottish schools, in which freedom and diversity in assessment methods and in regimes for remediation and differentiation were encouraged (Black 1986). A constructivist view of learning, arguing for the overt engagement of pupils in reflection on their own cognitive processes, leads directly to a need for formative assessment: *an active teaching role is advocated with a focus on monitoring and sustaining overt engagement of all students* (Tobin *et al* 1988a).

5.6. HOW CAN FORMATIVE ASSESSMENT BE DEVELOPED NOW?

The considerations presented above should have a profound effect on the curriculum and on assessment. The need for both to give a faithful picture of science and to develop productive styles of thinking are not served by current assessment practices and pressures; more varied, complex and educationally valid assessment methods need to be developed (Raizen 1990). The broader aims that science education should pursue are not at present widely supported by appropriate assessment methods.

Important general lessons arose from the work of the Science Assessment of Performance Unit (APU) over the period 1980 to 1990 (Black 1990):

- a wide range of methods must be used. The narrow range used in such exercises as the international evaluation studies and the USA National Assessment is unacceptable because it gives an unreliable picture of pupils' capabilities, and fails to recognize issues about learning and its assessment which are of central importance;
- tests limited to one or two questions for any one criterion, cannot give a reliable result, even for the average of a large group, let alone for one individual child. At age 13 in 1984 APU used 35 packages including 465 different questions in order to obtain reliable results. To take all of the practical and written tests in this set would have involved a pupil in about 35 hours of work (Johnson 1988);
- a teacher, who can record a pupil's performance over time and in several contexts, and who can discuss idiosyncratic answers in order to understand the thinking behind them, can build up a record of far better reliability than any external test can achieve. But in order to do this, teachers need support in the form of questions, procedures and in-service training.

All this should be of serious concern in any system where external tests are used to make important decisions about pupils' futures. In most public examination systems, not only have these problems not been solved; they have not even been clearly identified. The reason lies in the scale of resources devoted to appraising examinations. There have been very few systematic attempts to compare them with assessments which explore performance over a much wider range—of types of question and of assessment times—than public external examinations can allow. The expertise and intelligence built up in the APU Science research far exceeds that which those engaged in public examinations in science have been able to deploy in their task.

For the development of formative assessment, both as an end in its own right and a means to ensure that teachers play a greater role in summative assessment, several principles must be kept in mind. The first is to challenge the idea that the need to assess interferes with, and so harms, normal teaching and learning. How can either teacher or pupil proceed without checking that learning is effective, and without using immediate feedback to correct misconceptions and omissions? To teach without assessment feedback is to travel blind.

The second principle follows from the first. Assessment is not an extra to be attached to a piece of teaching like a barnacle to a ship's hull. It ought to be built in to the design of the teaching from the start. If such a design has clear aims, then pupils' work done while learning will naturally need to show how those aims are being achieved, and will thereby provide the needed formative assessment (see ASE 1990 for an excellent example of this approach).

A third principle is that pupils must become involved in their own assessment. If they can understand their own strengths and weaknesses, this will improve their learning by giving them more responsibility and control. Some of the advantages of such an approach have been shown in the development of records of achievement (Broadfoot *et al* 1990)

One reason why this vision of assessment is not easily grasped is that the model of assessment and testing which many of us have is the one established by our experience of external examinations, a point which Eric Rogers clearly recognized in his work to change teachers' views about testing. Because such examinations have such a strong influence on everyday teaching, assessment in classrooms cannot be improved in isolation from improvement of external testing. Just as the need for good formative assessment is undeniable, so too is the need for fair methods to provide certification for students. However, such information can only be fair to individuals, and useful to the public, if it is both valid and reliable.

It is quite difficult for external examinations to satisfy these conditions. This is because such examinations have to be short in time, and are set in artificial and stressful circumstances. If carried out with care, the information that teachers can assemble as part of formative assessment can escape many of these limitations. Thus an effective assessment policy must develop both formative assessment by teachers, and improved external examinations in which assessment by teachers should have a much greater part to play than in the past.

5.7.   A NEW WAVE?

There are several comprehensive guides to assessment for teachers which are well documented and which cover many of the issues discussed. In *Testing for Learning* Mitchell (1992) draws general guidance from developments in the USA, with examples from the sciences, and has an interesting section on getting students, parents and the community involved with assessment. Griffin and Nix (1991) give a particularly well-referenced account of the background to new approaches in assessment and reporting, with examples mainly from New Zealand and Australia. In Britain, Harlen's (1992) book *The Teaching of Science* has a full discussion of assessment relevant to primary science, whilst Connor (1991) deals with primary school needs across the whole curriculum.

Comprehensive guides for the development of teachers' assessment of science with practical examples, covering elementary and middle school science, have been published in the USA by the National Centre for Improving Science Education (Raizen *et al* 1989, 1990). They emphasize the need for a revolution in teachers' methods of assessment so as to reflect an improved curriculum. The use of a wide range of methods, involvement of pupils and the close linking of assessment and instruction are all stressed.

All of these share a vision for assessment, expressed recently by Perrone in words which are very similar to those which Eric Rogers often used.

In the end these fresh directions are not as complex as they appear. They call upon us to ask, in relation to purposes, what would cause us to say that our students are thinkers, readers, writers, or comprehenders of knowledge, and then to work out systematic processes to follow up such questions.

(Perrone 1991)

REFERENCES

A.S.E. 1990 *Teacher Assessment—Making it Work for the Primary School*. Association for Science Education, Hatfield UK

Bangert-Drowns R L, Kulik J A and Kulik C C 1991 Effects of frequent classroom testing *Journal of Educational Research* **85** (2) pp 89–99

Black H 1986 Assessment for learning *Assessing Educational Achievement* ed Nuttall D L (London: Falmer Press) pp 7–18

Black P J (ed) 1990 APU Science—the past and the future *School Science Review* **72** (258) pp 13–28

Black P J 1993 *Physics Examinations for University Entrance. An International Study* Produced under the auspices of the International Commission for Physics Education Document Series No 45 (Paris: UNESCO)

Blank R K and Dalkilic M 1992 *State Policies on Science and Mathematics Education Council of Chief State School Officers* Washington DC, USA

Block J H, Efthim H E and Burns R B 1989 *Building Effective Mastery Learning in Schools* (White Plains, NY: Longman)

Broadfoot P, James M, McMeeking S, Nuttall D and Stierer B 1990 Records of achievement: report of the national evaluation of pilot schemes *Assessment Debates* ed Horton T (London: Hodder and Stoughton) pp 87–103

Connor C 1991 *Assessment and Testing in the Primary School* (London: Falmer Press)

Dobson K ed 1985 *Revised Nuffield Advanced Science-Physics Examinations and Investigations* (London: Longman)

Dockrell B 1991 The effects of system wide testing: issues raised by a case study in a developing country *Studies in Educational Evaluation* **17** pp 41–49

Duschl R A and Wright E 1989 Teachers' decision making models for planning and teaching science *Journal of Research in Science Teaching* **26** (6) pp 467–501

Gifford B R and O'Connor M C ed 1992 *Changing Assessments: Alternative Views of Aptitude, Achievement and Instruction* (Boston and Dordrecht: Kluwer)

Griffin P and Nix P 1991 *Educational Assessment and Reporting; a New Approach* (Carrickville, New South Wales: Harcourt Brace)

Harlen W 1992 *The Teaching of Science* (London: David Fulton)

Harlen W and Qualter A 1991 Issues in SAT development and the practice of teacher assessment *Cambridge Journal of Education* **21** (2) pp 141–152

Herr N E 1992 A comparative analysis of the perceived influence of advanced placement and honors programs upon science instruction *Journal of Research in Science Teaching* **29** (5) pp 521–532

Hodson D 1986 The role of assessment in the 'Curriculum Cycle': a survey of science department practice *Research in Science and Technological Education* **4** (1) pp 7–17

Johnson S 1988 *National Assessment: the APU Science Approach* (London: Her Majesty's Stationery Office)

Mitchell R 1992 *Testing for Learning* (New York: Free Press Macmillan)

Mortimore P, Sammons P, Stoll L, and Ecob R 1988 *School Matters: the Junior Years* (Somerset: Open Books)

O'Connor M C 1992 Overview in B R Gifford and M C O'Connor (ed) *op cit* pp 9–36

Perrone V ed 1991 *Expanding Student Assessment* (Alexandria, Virginia, USA: Association for Supervision and Curriculum Development)

Raizen S A 1990 Assessment in science education *The Prices of Secrecy: the Social, Intellectual and Psychological Costs of Secrecy* ed Schwartz J L and Viator K L (Cambridge Mass., USA: E T C Harvard Graduate School of Education) pp 57–68

Raizen S A, Baron J B, Champagne A B, Haertel E, Mullis I V S and Oakes J 1989 *Assessment in Elementary School Science Education* (Washington: National Centre for Improving Science Education)

Raizen S A, Baron J B, Champagne A B, Haertel E, Mullis I V S and Oakes J 1990 *Assessment in Science Education: The Middle Years* (Washington: National Centre for Improving Science Education)

Resnick L R and Resnick D P 1992 *Assessing the Thinking Curriculum: New Tools for Educational Reform* pp 37–75 in Gifford and O'Connor *op cit*

Rogers E 1969 Examinations: powerful agents for good or ill in teaching *American Journal of Physics* **37** (10) pp 954–962

Rudman H C 1987 Testing and teaching: two sides of the same coin? *Studies in Educational Evaluation* **13** pp 73–90

Scott D 1991 Issues and themes: coursework and coursework assessment in the GCSE *Research Papers in Education* **6** (1) pp 3–19

Shepard L A 1992 Commentary: what policy makers who mandate tests should know about the new psychology of intellectual ability and learning, in Gifford and O'Connor *op cit* pp 301–328

Smith P S, Hounshell P B, Copolo C and Wilkerson S 1992 The impact of end-of-course testing in chemistry on curriculum and instruction *Science Education* **76** (5) pp 523–530

Sternberg R J 1992 CAT: a program of comprehensive abilities testing, in Gifford and O'Connor *op cit* pp 213–274

Tobin K and Garnett P 1988 Exemplary practice in science classrooms *Science Education* **72** (2) pp 197–208

Tobin K, Capie W and Bettencourt A 1988a Active teaching for higher cognitive learning in science *International Journal of Science Education* **10** (1) pp 17–27

Tobin K, Espinet M, Byrd S E and Adams A 1988b Perspectives of effective science learning *Science Education* **72** (4) pp 433–451

Wood R 1991 *Assessment and Testing: a Survey of Research* (Cambridge: Cambridge University Press)

Yager R E and McCormack A J 1989 Assessing teaching/learning successes in multiple domains of science and science education *Science Education* **73** (1) pp 45–58

Please do not think I am opposing tests and examinations. I am only dismayed by premature statistical enquiries that ignore serious aims. Examinations are powerful agents for harm or for good in programmes with new teaching aims. If teachers give the new suggested scheme a good trial but then set tests or exams which only match some older scheme of aims, the new scheme will face ruin: teachers will be guided away from the new suggestions by those exams, and so will students. Even if the test questions are set with new aims well in view, damage will come if the marking is done with the old spirit—for example a question that asks for constructive thinking but is marked with an item-list for diagram-neatness, definitions quoted, proper spelling, formulae stated, and all with neither place nor bonus for imagination.

Eric Rogers *Response to presentation of ICPE Medal* Trieste 1980

CHAPTER 6

EXAMINING INQUIRING MINDS?

Tae Ryu

6.1. THE INTRODUCTION OF NUFFIELD PHYSICS TO JAPAN

I vividly recall Eric Rogers in Edinburgh in 1975, demonstrating free fall by dropping two pieces of chalk. This was my first visit to Europe and my first international conference on physics education. Amongst the many international leaders in physics education whom I met there, Eric Rogers stands out as especially impressive. I have never forgotten his lecture, and since then have always used the same demonstration in introductory courses for non-science university students. I like the demonstration because it is so simple that students can do it themselves and so come to understand the work of Galileo.

A quarter of a century has passed since John Lewis introduced the Nuffield physics project to Japan, at a seminar in 1968 organized by Professor Akira Harashima, a director of the Physics Education Society of Japan. He told us the background philosophy of the project, showed us the teaching materials and demonstrated many experiments. I was excited by the main aim of the project, to foster inquiring minds in future citizens through students' own activities. It was a great pleasure for me, as an editor of the Journal of the Physics Education Society of Japan, to help in publishing a special issue on the Nuffield Physics Seminar. Participants in the seminar often quoted the proverb, 'I hear and I forget, I see and I remember, I do and I understand', and it became a watchword for science teachers who liked to do laboratory experiments.

We translated the Nuffield Teachers' Guides and other books into Japanese. At that time I was especially interested in the discussion of examinations written by Eric Rogers. To me it sounded humane, democratic and practical thinking about education. I particularly liked the questions asking pupils to reason about real phenomena, for example guessing the value of some quantity concerned with everyday physical phenomena and using it to solve a problem. It seemed really interesting to do, and to stimulate a scientific, inquiring attitude of mind.

6.2.   THE JAPANESE RESPONSE

In sharp contrast to the Nuffield Physics questions, I found that many of the physics questions in university entrance examinations in Japan were about artificial phenomena, testing only memory and facility in calculation. I wanted to change these tests, because they had so much influence on the teaching of physics in secondary schools. Students were trained to solve theoretical problems, but not to understand physics through experiment. We tried such new tests experimentally in part of the physics entrance examination for Sophia University, over a period of five years in the 1970s. However, the experiment was stopped by professors who were more interested in writing academic papers and not at all interested in education in schools. I was disappointed.

In order to promote a way of teaching physics which would make it understandable for future citizens, I organized meetings of teachers from high schools near Tokyo, who were interested in student-centred physics education. Because our Japanese system is much influenced by that of the USA and is quite different from that of the UK, I introduced Harvard Project Physics to these teachers, particularly the student activities and the independent study approach. From this the Group for Physics Education Research in Japan was born, which today has become the Association for Physics Education in Japan, which participates in international discussions and seminars. Some of its members are writing physics textbooks, under the control of the Ministry of Education. Now, as leaders of physics education in Japan, its members face the problems of education, especially those concerned with entrance examinations to universities.

As a matter of fact, the influence of entrance examinations on education in secondary schools has grown worse and worse, particularly since the advent of computer marking of answers, with national tests from 1978 onwards using only multiple-choice tests. Physics has become more and more unpopular as a subject, compared with chemistry or biology. Many physics teachers feel that there is a crisis in physics education, and worry not only about the future of physics, but also about the future of Japanese society.

I believe that we must change the system of entrance examinations from selection by university staff to certification by teachers, if we are to move towards student-centred education. This is very difficult to do, because many of the Japanese elite believe that the traditional selection system works efficiently as a way of selecting an elite, and so see it as important in maintaining the authority and privileges of the elite. It sometimes seems as if the implicit aims of education in Japan have not changed for a century: that is, to teach a Confucian ethic making children obedient citizens submitting to political and economic authority, rather than to foster humanity and creativity to produce rational and international citizens.

6.3. ANALYSIS OF ENTRANCE EXAMINATIONS

In Japan, there are two kinds of entrance examination; the National Centre Test and entrance examinations set by universities. The National Centre Test is required as the first stage of the entrance examination for all students entering any national public university, and for some faculties of private universities. In 1990, 60% of the 18 year old age group (1.1 million students) were candidates for university and college education. Two thirds of them went into higher education, with 431 000 students taking the National Centre Tests, 136 000 taking the physics test.

Essentially the whole test in physics consists of multiple-choice questions and short answer questions, in papers lasting from one to two and a half hours. By contrast the Nuffield physics A-level examination has several different kinds of paper, taking over seven hours, plus a two-week investigation.

Candidates, schools and universities are ranked by scores on these tests. The public very much sees the entrance examination as a selection device for the higher ranked universities, leading to privileged careers. The competition to enter prestigious universities becomes stronger and stronger. Parents urge their children to study hard to enter private secondary schools which prepare students for the entrance examinations of the highly ranked universities. The entrance examination for Tokyo University is particularly influential, it being the highest ranked university, which led me to do an analysis of the physics problems set from 1989 to 1992 in this examination and in the National Centre Test, comparing them with questions in the Nuffield A-level physics papers.

Figure 6.1 shows the coverage of topics by the Japanese tests, in 1989 and 1992, compared with the Nuffield A-level coverage in 1989. It is immediately clear that the Japanese tests cover fewer topics, and do not cover all the topics in the curriculum, with mechanics and electricity and magnetism having disproportionate weight. The National Centre test in 1992 had a little more variety than before, with a reform of the structure of the test which encouraged this.

The difficulty of the National Centre Tests has been widely criticized, particularly in 1989 when the mean score in physics was only 53%. In the next two years it rose to just over 70%, but fell again in 1992 to just below 60% with the introduction of new kinds of questions.

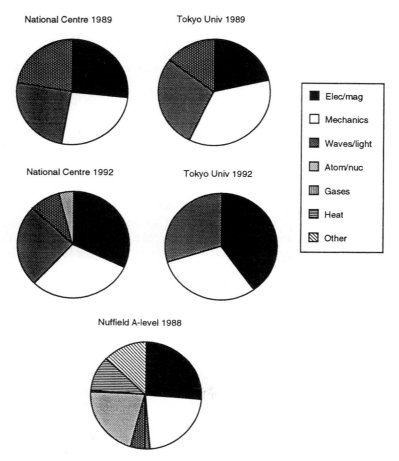

Figure 6.1. *Examinations: distribution of content*

The Nuffield examinations also require a greater variety of kinds of answer, as can be seen in figure 6.2. I analysed questions by the nature of the responses they required, putting the required responses in five groups:

qualitative,
numerical,
formula,
graph,
. diagram.

Figure 6.2 shows the weighting of marks given for each kind of response. In 1989 the National Centre Test was heavily weighted to answers requiring formulae, and of the small remainder only 5% went for qualitative answers, the rest being numerical. By 1992 the variety of answers rewarded had increased, in part perhaps because of an investigation into physics problems

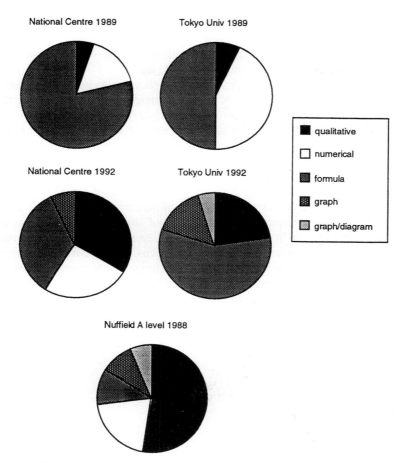

Figure 6.2.   *Examinations: distribution of types of answer required*

done by the Physics Education Society of Japan. Even so, formulae and numerical answers make up almost 60% of the total.

The tests for Tokyo University show only a slight movement towards greater variety. Formulae and numerical responses remain at least 80% of the total, the main change being the addition of some graph responses. The high proportion of formula responses in relation to numerical ones in both tests arises because of the highly theoretical, imaginary, artificial and unrealistic contexts in which problems are often set. Such problems require the ability to do algebra quickly, but nothing else. They are useless in the general education of a future citizen. As a result, non-science students hate physics, and the subject becomes unpopular, with the number of students enrolling for it falling drastically.

6.4. AIMS AND EXAMINATIONS

In 1972, the Japanese Ministry of Education set the following aims for physics in our national curriculum:

- to have pupils discover problems in natural phenomena and learn scientific methods through the process of studying them, thereby developing their creative ability;
- to have pupils understand the basic concepts of science and develop the ability to consider natural mechanisms and functions in a comprehensive and unified way;
- to have pupils cultivate a scientific way of viewing and thinking about phenomena in nature and to foster a scientific view of nature.

It is clear that university entrance examinations in physics are a very long way from measuring the achievement of these aims, and in no way encourage teachers to try to achieve them, even when they passionately wish to do so. If we want to teach to these aims, then we have to improve the examinations. We are doing what we can to get better examinations, and in doing so are learning much from Nuffield Physics. Above all, we think of what Eric Rogers would have advised us to do.

... Experiments... have shown that labs do no better than a class in teaching facts and principles of physics, or problem-solving skills—and labs cost far more in time and money. We do not use labs primarily for that; nor do we think students so stodgy that they understand only what they see and smell and touch. What we do expect from labs is a deeper and more lasting understanding of science, gained by working with one's own apparatus, by experiencing the frustrations and successes of experimenting, the delights and sorrows of a scientist. For understanding the interplay of experiment and theory, nothing can replace, for many students, the doing and discussing of their own experiments. And for all of us, delight comes strongest from the experiments we do ourselves. To sum up: this is where the student makes contact with adult science... I point to the analogy with learning a foreign language: labs are the study trip abroad.

Eric Rogers *Teaching Physics for the Inquiring Mind* p 98

CHAPTER 7

EXPERIMENTING WITH EXPERIMENTAL PHYSICS TEACHING

Anthony French

7.1. INTRODUCTION

It is generally taken for granted that experimental work is an essential and valuable part of any instruction in physics, and effort is continually being invested in its design. However, much of this effort is concerned with particular devices or experiments, whilst the general question of the purpose and value of experimental work is less often discussed.

Experimental work, as a component of education in physics, is not much more than a hundred years old. One of the first to advocate it was James Clerk Maxwell. We think of him primarily as a great theorist, but it is good to remember that, in his inaugural lecture in October, 1871, as the first Cavendish Professor of Physics at Cambridge University, his chief topic was the teaching of experimental physics. He was heartily in favour of it. Not everyone agreed: one of my favourite quotations (although I reject its sentiments) is from Isaac Todhunter, a Cambridge mathematician. He was first in the Mathematical Tripos at Cambridge in 1848. (Maxwell came second in this examination in 1854!) In 1873 Todhunter wrote.

We assert that if the resistance of the air be withdrawn a sovereign and a feather will fall through equal spaces in equal times. Very great credit is due to the person who first imagined the well-known experiment to illustrate this; but it is not obvious what is the special benefit now gained by seeing a lecturer repeat the process. It may be said that a boy [there were of course no female students!] *takes more interest in the matter by seeing for himself... It may be said that the fact makes a stronger impression on the boy through the medium of his sight, that he believes it the more confidently. I say that this should not be the case. If he does*

83

not believe the statements of his tutor—probably a clergyman of mature knowledge, recognized ability, and blameless character—his suspicion is irrational, and manifests a want of the power of appreciating evidence, a want fatal to his success in that branch of science which he is supposed to be cultivating.

(Weber 1973)

The Maxwell attitude prevailed, and practical physics proceeded to become a standard component of all physics instruction (Phillips 1981).

7.2.   WHY HAVE PHYSICS TEACHING LABORATORIES?

What specific purpose or purposes should a physics teaching laboratory serve? Anyone who teaches practical physics is aware that the possible roles of practical work are various. Different authors categorize these roles in different ways. One of the best discussions I know is that of Neher (1962) who identifies the following reasons (among others) for doing laboratory work:

- most people learn with their hands as well as with their heads (e.g. through a first-hand experience of the magnitudes of different forces);
- a good laboratory should develop the student's experimental ability (e.g. judging what is important, and choosing the best order of procedure);
- many experiments illustrate physical principles that are very difficult to grasp from classroom work alone (e.g. the subtleties of electrostatics);
- part of education in physics is a training in skills (e.g. such simple things as making adjustments and accurately reading scales;
- there are certain ways of doing things that are best taught in a laboratory (e.g. how to keep a notebook, how to tabulate and analyse data, how to plot relevant and appropriate graphs).

Underlying all this is the fact that physics is a particular way of learning about nature. To succeed in this quest, it is necessary to be selective and systematic in one's experiments. We cannot expect the beginning student to develop proficiency in this endeavour in one large jump, but we can try to acquaint him or her with the various ingredients that make for success.

7.3.   SOME AWKWARD REALITIES

Far too often, the practical class fails to do what we intended. In recent years, especially, there have been some sobering studies of such matters. They can be roughly divided into two categories (although they are closely related).

(1) The ineffectiveness of practical work in general, as an educational tool;

(2) the lack of understanding, or the sheer misunderstanding, that may remain after a student has completed an experiment designed to illuminate some particular piece of physics.

The culprit is often the type of 'cookbook' laboratory in which a student, following a set of detailed instructions, carries out a set of measurements and analyses the results—in other words, the type of exercise found in introductory physics labs all over the world.

With respect to point (1), there are some very discouraging results. Sanborn Brown of MIT (aided by J G King) found from a survey that the experiments done by freshmen at MIT were similar to those many had done in high school (Brown 1958). There was no significant difference in lab performance at MIT between those who had done similar experiments in high school and those who had not. Even more disconcerting, students were shown pictures of 25 pieces of apparatus used in experiments they had done in high school, and were asked to identify the apparatus in question and what physical quantity was measured with it. Only about 40% of the answers were correct. The previous lab experience had scarcely laid a glove on more than half of the students!

Another study by Toothacker (1983) questioned the hallowed assumption, justifying the expenditure of large amounts of time and money, that laboratory work strengthens or deepens a student's understanding of theory. He cites several studies indicating that performance in written examinations is little affected. Long, McLaughlin and Bloom (1986) concluded that laboratory work had a small positive influence (about a third of a letter grade, on the average) for students of middling ability, but had no perceptible effect on the performance of the best and the weakest students.

Even more shocking are the results of careful, detailed studies of point (2), based on one-to-one discussions between a student and an interviewer in front of a piece of apparatus. McDermott (1991) gives a recent and comprehensive account of this work. An earlier article (Goldberg and McDermott 1986) begins.

Ask a student who has just finished studying geometrical optics where the image of an object placed in front of a plane mirror is located, and the student is likely to reply without hesitation that the image is the same distance behind the mirror as the object is in front. Moreover, the student can probably produce a ray diagram to justify this answer. However, pose a slightly more complicated question that may not have been specifically addressed, and the outcome is likely to be very different. For example, if you ask a student, who has not been expressly taught the answer, whether his distance from a small mirror would affect the amount he could see of his own image, the response is almost certain to be incorrect. Given time and encouragement to reconsider, the student very probably will not even be able to draw a ray diagram that might help answer the question.

The article goes on to document the extent of this lack of understanding of image formation by plane mirrors—a process that most physics teachers might regard as simple and scarcely in need of elaboration. The same authors provide similarly disturbing data on the more complicated phenomena of image formation by converging lenses or concave mirrors (Goldberg and McDermott 1987). One particularly instructive question was to ask what would happen to the real image formed by a convex lens if the upper half of the lens were covered. A majority said that half of the image would disappear, and even justified this with a ray diagram.

Such studies tell us an important lesson. This is that when students perform an experiment by following a set of 'cookbook' instructions they may, superficially, appear to have performed satisfactorily, but they may have understood very little about the underlying physics, and it takes extra probing to expose this fact. They may have derived certain benefits—the ability to interpret the instructions, the development of some manipulative skills, the act of becoming acquainted with the appearance of some piece of apparatus—but they are not learning at a deeper level, in a way that will enable them to apply what they have learned in a different or wider context.

7.4.   ROADS TO IMPROVEMENT

What can be done to improve the situation? The basic problem to be faced is that, for many students, the lab work in an introductory physics course is a largely meaningless exercise, devoid of intrinsic interest. The student goes through a set of prescribed exercises, and comes out little the wiser. An extreme solution is to abolish the lab altogether; this was in fact recommended by Toothacker (1983) so far as introductory physics is concerned, and I may add that this was actually done by MIT in 1964, as a result of disenchantment with the labs then in existence! But this is running away from the problem. Our students need to come to grips with the real physical world, not to base their knowledge of it on lectures (even with demonstrations) and textbooks.

Could one do a follow-up interview on every experiment, probing each student's understanding of it? In most places that would be an unattainable luxury. A carefully organized laboratory incorporating some aspects of this strategy has, however, been successfully tried at the University of California, Berkeley (Reif and St. John 1979). Another valuable approach is to get away from labs that merely involve the verification of already well known results. Chimino and Hoyer (1985) have introduced audio-tutorial materials (35 mm slides and cassette tapes) to provide a link between labs and theoretical parts of a course. Menzie's (1970) thoughtful question, 'What is it that the laboratory can best contribute to a student's education?' can usefully be turned around: 'What aspects of a student's education in physics do we, as teachers, wish to foster through experimental work?' Any answer to the question must be a matter of personal choice—which brings me to the particular response that my colleague John King and I have been developing and using at MIT for the past few years.

7.5.   A PHYSICS COURSE WITH TAKE-HOME EXPERIMENTS

In creating our new course at MIT, John King and I have firmly come down on the side of process rather than specific subject-matter as the most enduring and transferable aspects of an introductory physics course (for arguments see Michels 1957 and Fowler 1969). Here I will concentrate on the experimental component; other aspects of the course are described elsewhere (French 1989).

A very novel feature of the course is that all the experimental work is based on kits of simple equipment, which students take home or to their dormitory, where they assemble the apparatus and make measurements in their own time, normally in pairs. Written instructions are provided, but with room for individual ingenuity. Most of the experiments involve a fair amount of assembly and construction for which each student is given a tool-kit containing such items as a soldering-iron, wire-strippers, pliers, screwdrivers, an inexpensive electrical multi-meter, and a 12 V mains transformer.

The principal purpose of the course is to let students experience what experimental physics is actually like. Things do not necessarily work well right away; one has to cope with the frustration! We emphasize that the progress of the work should be recorded as it happens—sketches and brief descriptions of set-ups, tabulations of raw data, appropriate graphs, and theoretical analysis. We insist that we do not want carefully organized reports written after the event; we want an honest picture of what happened when, including errors and false starts. We try to make clear that the value of the exercise lies as much in the successful carrying out of the task as in a satisfactory result.

Electrical measuring techniques play an important role throughout, as they do in real experimental work in any field. Just because an experiment is in mechanics, we do not feel inhibited from asking the student to assemble a simple circuit with which to make measurements, even though theoretical work in electricity and magnetism still lies ahead.

We place great emphasis on the graphical analysis of data, and on the value of finding ways to get linear graphs given some theoretical expectation. We stress that no such relationship is necessarily valid over an arbitrarily wide range, and that deviations from it can be significant and instructive, and so should be looked for and understood (for example, the effect of the finite size of light source and detector in an experiment to study the inverse-square law).

The experiments themselves do not constitute a frozen or complete list. The design of individual experiments is constantly being reviewed, and the total number of experiments is added to as circumstances permit. The goal is to have a repertoire from which instructors can make their own choice; there is nothing sacred about any particular experiment—although we do have some favourites!

These are some of the general principles underlying our programme, but the best way of illustrating them is to describe some specific examples; that will be the purpose of the next section.

7.6.   A SELECTION OF THE TAKE-HOME EXPERIMENTS

7.6.1.   Range-Finder

This very simple experiment requires almost nothing in the way of construction or assembly, and has served as the first experiment in the course. It calls for observations to be made outdoors, reminding students that physics is not confined to the classroom. It can easily be adapted to the local landscape anywhere.

As performed at MIT, the main object is to measure the 600m width of the Charles River. Students are provided with an enlarged photocopy of a quadrant of a protractor, a drinking straw to use as a sighting device, a long drawing pin, and a rubber band. The pin pivots the straw at the centre of the quadrant, and the band holds it in place. Starting at an arbitrary point on the river bank, students sight on some object at the far side of the river, and measure the angle α between the river bank and the line of sight to the object, repeating this for various distances x along the bank from the starting point. Instead of calculating the river width D for each triangulation (as our students tend to do) they are asked to find a best value for the width from a graph (the graph of $\cot \alpha$ against x has slope of magnitude $1/D$). They may also estimate the height of buildings across the river. This may seem an excessively simple experiment for freshmen at a prestigious American university, but I have to say that many of our students do not come well equipped to do a good job on it.

7.6.2.   Measuring light intensities

This experiment concerns two of the most pervasive functional relationships between physical quantities—the inverse square law and the exponential. It uses a 15-inch 'optical bench' made of Dexion, an automobile tail-lamp bulb and a photo-resistor, with absorption filters made from layers of grey translucent plastic tape.

Students measure how the resistance of the photo-resistor varies with the number of layers of filter material for fixed light source and detector, and how it varies with the distance between source and detector, with the filters removed. One way of analysing the data is to assume that the resistance varies inversely with light intensity, and thus check the exponential variation of intensity with the number of filter layers, and the inverse square variation of intensity with distance from the source. Or they can take the inverse-square law as given, using it to calibrate the photo-resistor and then analysing the absorption data.

With the photo-resistor sufficiently close to the lamp, one can also measure a luminous intensity about equal to that of direct sunlight. Taking the distance of the Sun as known, students can estimate its total power output.

7.6.3. *Falling objects*

In the way we measure the acceleration of free fall, students have to construct their own electrical timing circuit.

The experiment uses the Dexion strip, now mounted vertically. A steel ball-bearing is taped to a piece of paper-clip wire, and clamped by an alligator clip at some initial height. Opening the clip to release the ball removes a short circuit across a capacitor, which begins to charge up from a low-voltage dc source, which could be a 9 V battery. When the ball falls into a small paper cup clipped lower down, the charging circuit is interrupted. The time of fall can then be inferred from the voltage developed across the capacitor. The time constant of the circuit is long enough (about 10 s) for the capacitor voltage to be essentially proportional to the time of fall (up to about 200 ms). A good straight-line graph can be obtained of the square of the time (or voltage) against the distance of fall, and the value of g can be obtained with about 5% accuracy.

7.6.4. *Electronic power supplies*

Two important ingredients of the take-home kit are parts for a regulated low-voltage power supply and for an associated high-voltage supply. Both are made by the students using standard 'breadboard' techniques. The low-voltage supply, giving from about 2 to 12 V at currents up to about 1.5 A dc, is run from the 12 V transformer. It also drives the high-voltage supply, which gives up to about 1 kV at a current of less than 1 mA, so that it is completely safe. The high-voltage transformer is made from a ferrite core inductor (several hundred turns) wound with a small number of turns of enamel-covered wire.

7.6.5. *Centripetal force*

In this very simple experiment, a small nut attached to a rubber band is whirled around at high speed by a small dc motor, originally designed to advance film in disc cameras, driven by the low-voltage supply. To obtain the lowest speeds, a dry cell is connected in opposition to the supply, whose minimum output is about 2 V. The light bulb (previously used in the light intensity experiment) helps to hold the speed of the motor steady, by virtue of its large variation of resistance with current.

The motion is strobed with an LED driven from the 12 V transformer, giving 60 flashes per second, to select several well defined rates of rotation. The radius of the circular path of the nut is judged by eye by holding a ruler above the plane of rotation and looking down from above. Centripetal accelerations obtainable in this experiment are impressively high—well over 100 g.

A static experiment is performed to measure the length of the rubber band as a function of tension, using 1-cent coins in a paper cup as loads, to give an empirical calibration of the non-linear relationship.

7.6.6.   Electrostatic force

This particularly simple experiment measures the voltage from the high-voltage supply across two horizontal plates needed to lift a small piece of foil in the field between them. Knowing the weight of the foil, the dielectric constant of free space can be found. The plates, (used also in an experiment on angular momentum), are two 2.5 inch diameter metal washers placed horizontally and separated by small pieces of insulator. Pieces of aluminium foil are cut from kitchen wrap, and folded so as to make small squares with one, two or three thicknesses of foil. Each square in turn is placed flat on the lower washer. There is a critical voltage at which the square of foil is just lifted by the electrostatic force. The force per unit area is proportional to the square of this voltage, so that knowing the density of aluminium, and the spacing of the plates, the dielectric constant of free space can be found from a graph of the square of the critical voltage against the number of layers of foil.

7.6.7.   Magnetic force

This experiment measures the force between two closely-spaced, flat, circular coils carrying the same current. One coil sits on a horizontal surface; the other is mounted on the underside of one arm of a sensitive balance, made of a strip of light-weight foam-core plastic board, pivoted by means of long pins on a pair of corner brackets. The beam is balanced, with no current, with the help of one or two coins. Measurements of the force between the coils as a function of current, with a fixed small spacing between the coils, are taken using weights made from aluminium foil. A conceptually more tricky alternative to using the low-voltage power supply is to use ac from the 12 V transformer and to vary the current with the help of two tail-light bulbs, each

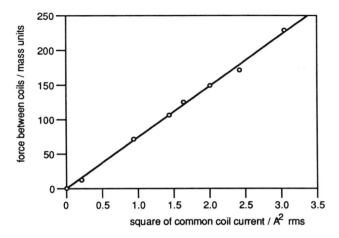

Figure 7.1. *Representative current balance data*

with a high-resistance and a low-resistance filament, connected in various ways, to give a nicely spaced set of current values. Deciding how to do this is an interesting side-exercise in itself, and this is another example of using the same piece of equipment for several different experiments. Figure 7.1 shows the kind of results one can get.

7.6.8. An electronic electrometer

Quantitative experiments in electrostatics are usually hard to do. But they become simple with the electrometer that John King designed for our course. The basis of it is an operational amplifier with a very high intrinsic gain that can be used, with negative feedback, to make an amplifier of moderate gain with an extremely high input impedance (figure 7.2). A charge placed on the input wire will remain with almost no leakage, and the amplifier output gives a measure of the amount of this charge. With this device it is possible, for example, to explore the distribution of charge on a charged, insulated conductor, or study the transfer of precisely defined amounts of charge from a high-voltage source via a capacitor.

7.6.9. Producing and detecting microwaves

Some astonishingly simple additions to the items they have already constructed allow students to study the production and detection of electromagnetic waves. To generate the waves, the high-voltage supply is connected across a small adjustable spark-gap between the heads of two drawing pins. A resonant dipole antenna, made of short lengths of copper wire connected to the two sides of the spark gap, radiates at a wavelength

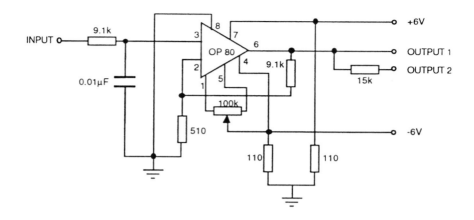

Figure 7.2.   *Electrometer circuit*

of about 15 cm, the waves being detected with a similar tuned dipole and
a diode. The rectified signal goes to the electrometer, with its negative
feedback reduced so that its gain is increased to about 100. This is essentially
a miniaturized and modernized version of the way Heinrich Hertz himself
first demonstrated the existence and the properties of electromagnetic waves
in 1888 (Hertz: translated Jones 1972).

The linear polarization of the waves can be detected by rotating
the receiving antenna, or using a parallel-wire polarizer—the microwave
equivalent of a piece of Polaroid sheet. With a reflector made of a sheet
of aluminium foil one can get standing waves, and deduce the wavelength
from the distance between successive maxima.

This is probably the most exciting, but also the most difficult, of
our experiments. Its difficulty lies in the erratic behaviour of the spark
transmitter, which does not always oblige by producing a strong output
of 15 cm waves. It demands patience, and perhaps a bit of luck, to achieve
really good conditions. But the effort is worth it!

Our complete menu includes additional experiments in mechanics and
electricity, and also several in the areas of mechanical and thermal energy,
acoustics and physical optics. The last are based on the use of a photo-
multiplier tube to study interference and diffraction patterns, spectra and,
finally, to observe the photoelectric effect, show the linear dependence of
photoelectron energy on frequency, and make a rough determination of the
Planck constant.

7.7.   AN INFORMAL EVALUATION OF THE PROGRAMME

It is hard for innovators to be objective about their products. We, of course, believe that the style and the content of our programme represent a substantial improvement over traditional experiments performed in a laboratory within designated hours, using prefabricated equipment. There is no doubt that many of our students feel the same, and have fun tinkering with the apparatus they have themselves constructed. It can give a student a sense of pride to take home and display to friends a high-voltage power supply (for example) he or she has constructed.

Some students are not so charmed, becoming exasperated when the apparatus does not do what it is supposed to do, or resenting the amount of time it takes to complete an experiment. Some students and colleagues legitimately criticize the absence of sophisticated equipment and the relatively low accuracy of most of our measurements. We certainly do not wish to encourage our students to be sloppy, but the other side of the coin is to discover how good a measurement one may be able to make with very simple and inexpensive apparatus—enough to give an excellent quantitative appreciation of the phenomenon being studied. In our view, refinement can come later. But that, I admit, is a point of view. Some students tend to look down their noses at our kits, with their unimpressive assortments of bits and pieces. They would like to get their hands on elegant engineered devices and enjoy the sense of power that comes from manipulating them. If so, they are likely to dismiss our approach—although I still think it would be good for them! The one thing in this category that we would dearly like to provide for our students is a simple oscilloscope. We use oscilloscopes freely in lecture demonstrations in which the current week's experiment is introduced to them, but that is not the same thing.

A big question about any new programme is, of course, whether it can be successfully adopted by others. At MIT, our course has been taught only by its inventors and developers. However Professor Jerome Pine at the California Institute of Technology decided that he would like to try using the experiments in the electricity and magnetism part of the course there. CalTech is much smaller than MIT, and Professor Pine adopted our take-home experiments as the form of the laboratory for the whole freshman class. In a letter he wrote about it, he said:

Our evaluation of the effects of these labs is quite subjective, but we feel they have been very successful. Student performance on the lab quizzes shows an amazing amount of new knowledge, compared with what they came to CalTech with... For a sizeable number of students—perhaps 20%—the labs open up a new and exciting view of the pursuit of science. Colleagues report that they see a real improvement in students' ability in later lab courses.

We take this as encouraging evidence that this way of teaching experimental physics is transferable. It must be admitted that the programme needs a considerable investment of human effort. People must be available who can give help and advice to students who are having trouble with making their experiments work. And the logistics of assembling and distributing the kits has been a considerable challenge, though with experience this is getting easier. Above all, it requires instructors who are genuinely enthusiastic and who themselves love playing with the experiments. But is not such enthusiasm needed for any effective teaching?

7.8.   CONCLUDING REMARKS

As mentioned before, we do not believe that any one way of teaching experimental physics has all the answers. Any choice is bound to involve not only personal taste, but also a balancing of competing priorities. We do, however, take satisfaction in the fact that our approach echoes the views expressed by Clerk Maxwell in his inaugural lecture. He classified laboratory work into what he called Experiments of Illustration and Experiments of Research, and of the former type he said:

> *The simpler the materials of an illustrative experiment, and the more familiar they are to the student, the more thoroughly he is likely to acquire the idea which it is meant to illustrate. The educational value of such experiments is often inversely proportional to the complexity of the apparatus. The student who uses home-made apparatus, which is always going wrong, often learns more than one who has the use of carefully adjusted instruments, in which he is apt to trust, and which he dares not to take to pieces.*

There is also the mundane consideration that most of the materials for our experiments are very cheap. This should make it possible for programmes of lab work based on our experiments, or others like them, to be established almost anywhere. Only the future will tell if the take-home kit idea will catch on, but our hope is that this kind of approach to introductory physics labs will find favour in many places, including secondary schools. In the meantime, we ourselves plan to continue refining and adding to the experiments for our own use, and we welcome enquiries about them. One thing is sure: the vitality of physics education depends on constant re-examination and change, and we ourselves have had a wonderful time making our own particular contribution. We are grateful to MIT for providing financial support, for looking with a friendly eye on these efforts, and for allowing us to use freshmen as volunteers in trying them out.

ACKNOWLEDGMENTS

Philip and Phylis Morrison played an important role in developing the electricity and magnetism part of the course, and have co-authored with John King a manual about the associated experiments.

This article is a condensed version of a talk given at the Third Caribbean Conference in Physics, University of the West Indies, St. Augustine, Trinidad, February 1992.

REFERENCES

Brown S C 1958 *Am. J. Phys.* **26** 334–337
Chimino D F and Hoyer R R 1983 *Am. J. Phys.* 51, 44–48; 1985 *Am. J. Phys.* **53** 76–80
Fowler J M 1969 *Am. J. Phys.* **37** 1194–1200
French A P 1988 *Am. J. Phys.* **56** 110–113 and 1989 *Am. J. Phys* **57** 587–592
Goldberg F M and McDermott L C 1986 *Phys. Teach.* **24** 472–480
Goldberg F M and McDermott L C 1987 *Am. J. Phys.* **55** 108–119
Hertz H (translated Jones D E) 1962 *Electric Waves* (New York: Dover Publications) Reprint
Long D D, McLaughlin G W and Bloom A M 1986 *Am. J. Phys.* **54** 122–5
McDermott L C 1991 *Am. J. Phys.* **59** 301–15
Menzie J C 1970 *Am. J. Phys.* **38** 1121–27
Michels W C 1957 *Am. J. Phys.* **25** 82–8
Neher H V 1962 *Am. J. Phys.* **30** 186–90
Phillips M 1981 *Am. J. Phys.* **49** 522–7
Reif F and St. John M 1979 *Am. J. Phys.* **47** 950–7
Toothacker W S 1983 *Am. J. Phys.* **51** 516–20
Weber R L ed 1973 *A Random Walk in Science* (London: Institute of Physics Publishing) p 154
ZAP! *A Hands-on Introduction to Electricity and Magnetism* (preliminary edition) 1991 (Burlington, NC: Neal Patterson Publications)

Demonstrations also brighten the course with entertainment; but that is a poor objective. The common comment that a lecturer should be a showman is misleading: he needs to discuss science, and show science, not advertise science by a circus performance. However the course is run, some demonstrations seem to me to be essential, to emphasize the authority of experiment.

Eric Rogers *Teaching Physics for the Inquiring Mind* p 19

CHAPTER 8

DEMONSTRATION IN TEACHING AND IN POPULARIZING PHYSICS

Charles Taylor

8.1. INTRODUCTION

Lecture demonstrations are held by most physicists to be good in principle but too much trouble in practice. Eric Rogers was one of the relatively few physicists prepared to take the time and trouble to demonstrate often and well. Sir Lawrence Bragg, writing about the famous Christmas Lectures at the Royal Institution in London, says:

It is surprising how often people in all walks of life own that their interest in science was first aroused by attending one of these courses when they were young, and in recalling their impressions they almost invariably say not 'we were told' but 'we were shown' this or that.

I myself find it difficult to recall any of the lectures, many by distinguished lecturers, which I attended as an undergraduate, except for those on sound by Dr Alexander Wood and on optics by Sir Lawrence Bragg. Illustrated by superb demonstrations, their lectures remain as clearly in my mind as if they were given yesterday. When Alexander Wood showed us the Miller Phonodiek to display wave-forms, his flair for the dramatic led him to use the walls of the lecture theatre as a screen so that the whole class was surrounded by wave traces. I can even remember the music that he used to drive the device: the Rachmaninoff Prelude in G Minor. Both music and wave-forms are still engraved on my mind!

8.2. DEMONSTRATION IN SCHOOL

What do we mean when we talk about demonstration at school level? It could be taken to mean anything that goes beyond just speaking and writing. It could include hand waving to describe a spiral staircase, slides, video recordings, analogue experiments used to illustrate a difficult point, or actual experiments. A typical useful analogue experiment is the use of a ripple tank to illustrate reverberation of sound in a room, or interference in optics. A real experiment might be the famous illustration of the speed of molecular motion by admitting liquid bromine into an evacuated tube; perfect for demonstration because it makes just one visual point, besides being too dangerous for a class to perform for themselves.

Is there any case for demonstrations that could have been done by the pupils themselves? I think that there is, when it provides a peg on to which the memory of a principle can be hitched. Eric Rogers believed in elegantly simple experiments to be used at the time a principle was being discussed. Two which I remember particularly were demonstrated by him at the ICPE conference held in Edinburgh in 1975 (transcript in this volume). In one, he projected a drawing of a parabola on to a screen, and threw a coin so that its shadow exactly followed the parabolic path on the screen. The second involved two lengths of string to which weights were attached: the first had weights attached at equal intervals. When the upper end was released the time intervals between the thuds as each small weight hit the floor were very obviously decreasing. This demonstrated both the acceleration of falling bodies and the ability of the ear-brain system to discriminate small time intervals. The experiment was repeated with the second string. But now the weights were placed at distances which increased following a square law. When this string was released it was very clear that the thuds were equidistant in time.

8.3.   DEMONSTRATION FOR PUBLIC UNDERSTANDING

One of the all-time stars in the use of demonstration for promoting the public understanding of science was Michael Faraday who, in 1826, founded both the Friday Evening Discourses and the Children's Christmas Lectures which, of course, still continue. Many of the demonstrations he developed are still in use at the Royal Institution and it is one of the great thrills of lecturing there to be able to work with some of his actual apparatus. He offers some gems of advice; here are a few examples:

... I disapprove of long lectures. One hour is enough for anyone, and they should not be allowed to exceed that time.

In lectures, and more particularly in experimental ones, it will at times happen that accidents or other incommoding circumstances will take place. On these occasions an apology is sometimes necessary, but not always. I would wish apologies to be made as seldom as possible, and generally only when the inconvenience extends to the company. I have several times seen the attention of by far the greater part of an audience called to an error by the apology which followed it.

If the lecture table appears crowded, if the lecturer (hid by his apparatus) is invisible, if things appear crooked, or aside, or unequal, or if some are out of sight without particular reason, the lecturer is considered (and with reason) as an awkward contriver and a bungler.

These and many more pearls of wisdom, together with some equally valuable ideas from Sir Lawrence Bragg are contained in a small volume, *Advice to Lecturers* published for the Royal Institution.

Twenty or thirty years ago, except for such places as the Royal Institution and the Lawrence Hall of Science, public lectures for increasing public understanding of science were relatively rare and scientists who gave them were sometimes thought by their colleagues to be wasting their time and talents. A good demonstration lecture should be a dramatic performance, and the lecturer must be at least a performer, if not an actor. Perhaps that is why some scientists have felt in the past that there is something not quite respectable about such an activity. Thankfully this attitude is becoming a thing of the past. In the UK the Royal Society's Committee on the Public Understanding of Science, set up in 1985 as a result of the lead set by Lord Porter (one of Michael Faraday's successors as Director of the Royal Institution, then the President of the Royal Society), has done a great deal to change this image.

Eric Rogers certainly believed in dramatic presentation. In his series of Christmas lectures at the Royal Institution, he wanted to demonstrate the way in which surface tension controls the shape of a drop of water. He had a thin rubber membrane stretched across one end of a metal cylinder to make a tank about a metre in diameter and about half a metre deep. The whole was suspended from the roof of the lecture theatre and he then proceeded to mount a step ladder and to pour in buckets of water, handed to him by an assistant. As more and more water was poured in, the membrane assumed the characteristic drop shape. He knew that it took fourteen buckets of water to produce the drop and that adding the fifteenth would usually cause the membrane to rupture and deposit well over a hundred litres of water into a child's paddling pool placed underneath to receive it. To make sure that it really would rupture, an old razor blade was secreted in the fifteenth bucket of water. It only needed a slight nick in the membrane for the whole to tear apart. Drama was created by the build up of tension both in the membrane and in the audience, and few people present will forget just how a drop of water is formed!

8.4.   WHAT OF THE FUTURE?

Many people are convinced that the ready availability of video techniques, both tape and disk, is very likely to make live lecture demonstration obsolete. Their use is said to be much more efficient and to enable much more effective use of a lecturer's time to be made than with live demonstrations. The demonstrations for a video recording can be set up with plenty of time in hand, retakes can be made until everything works perfectly, and the same material can be used with vastly increased size of audience.

Against this, I think that some remarks made by Gerald Holton some twenty years ago need to be weighed in the balance:

> ... *there are a number of important values which, it appears to me, are conveyed in the interaction between the live lecturer and the student in the same room, particularly through the scientific demonstration; values which depend on communication being initiated by an actual person rather than by a surrogate.*

Further, when demonstrations are honed to perfection on a video presentation, the horribly wrong idea that experiments necessarily always work may be passed on. In real science one learns more from an experiment that goes wrong than from one that works the first time, if indeed there is ever one that does. Certainly the demonstration that goes wrong is the one remembered best by the audience!

...scientists grew much more careful about calling their models true. Their students were ready to believe the model tells why, but the best thinkers were once again stepping down from trying to say WHY to saying WHAT. That is an old lesson, never fully learned—if it were, scientists would lapse from credulous glorying into over-cautious denial—but a lesson to be applied again and again. As a good scientist you should be suspicious of models, particularly if they are treated as true descriptions. Yet you should not despise models as childish. They play an essential part in the human mind's method of perceiving and learning. When our senses tell us about something quite new our first mental move is to find something familiar that it reminds us of. We attach the old familiar label very strongly to the new thing, and only very slowly change to a new view.

Eric Rogers *Physics for the Inquiring Mind* p 718

CHAPTER 9

MODELLING CLAY FOR COMPUTERS

Jon Ogborn

9.1. MAKING WORLDS

To create a world on the computer, and to watch it evolve, is a remarkable experience. It can teach one what it means to have a model of reality, which is to say what it is to think. It can show both how good and how bad such models can be. And by becoming a game played for its own sake it can be a beginning of purely theoretical thinking about forms. The microcomputer brings something of this within the reach of many pupils and teachers.

The obvious way to make worlds with the computer is to make worlds built of equations relating variables. What we need are ways to make worlds which are simpler than this and which are closer to young people's understandings of the real world.

9.2. MODELLING WITHOUT MATHEMATICS

Can we do modelling without mathematics? Consider what is needed to make a mathematical model.

(1) imagining the world constituted of *variables*,

(2) conceiving physical relations as *mathematical relations* between variables,

(3) giving appropriate *values* to variables,

(4) seeing a model as a structure with *possibilities*.

The first step is perhaps the hardest. We have become so used to imagining the world as able to be analysed through the interaction of quantitative variables that we forget what a huge step in imagination this is. There is good evidence, supported by common sense observation, that young students see the world as built of *objects and events*, not as built of variables.

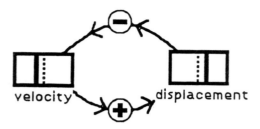

Figure 9.1. *An oscillator in IQON*

We have built, and tested with students in the age range 12–14 years, a modelling programme which focuses just on imagining variables and the connections between them, without having to specify the form of mathematical relations. It was developed in the project *Tools for Exploratory Learning* in association with Joan Bliss, Rob Miller, Jonathan Briggs, Derek Brough, John Turner, Harvey Mellar, Dick Boohan, Tim Brosnan, Babis Sakonidis, Caroline Nash and Cathy Rodgers.

The modelling system is called IQON (Interacting Quantities Omitting Numbers). In IQON one creates and names variables, and links them together graphically. The best introduction is by example: figure 9.1 shows what an oscillator looks like when expressed in IQON.

A positive velocity progressively increases the displacement, through the 'plus' link. But a positive displacement progressively *decreases* the velocity, through the action of a spring, represented via the 'minus' link. The result is that the system oscillates: negative feedback gives oscillation. What is shown in figure 9.1 is all that the user has to do: to create and name two variables and to link them as shown. No equations are written at all.

However, IQON is also intended for thinking about systems where we have much vaguer ideas about quantities and their relationships. Consider the quality of a conference. We may imagine that much depends on the quality of the workshops. If that is high, the participants become happier and happier as the week goes by. But if they are happy they may perhaps participate more actively in workshops, so that the quality of workshops itself increases. Figure 9.2 shows this idea expressed in IQON.

This model is overly optimistic. It contains positive feedback, so that if, as in figure 9.2(a), the quality of workshops is somehow increased by a small amount, then after some time all the variables are driven to their positive limits. It does not matter whether the model is correct; what matters is that such effects are possible and will certainly arise in some cases, whatever the details of the system. An increase in global temperature causing the melting of polar ice, which, by reducing reflectivity increases the energy absorbed

(a) initial setting

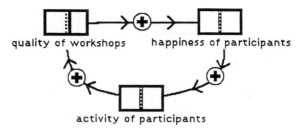

(b) positive feedback causes runaway

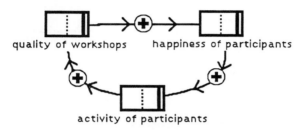

Figure 9.2. *An IQON model for success of workshops*

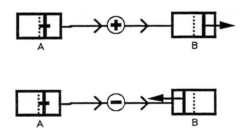

Figure 9.3. *Behaviour of linked variables in IQON*

from the Sun and so leads to a further increase of global temperature, is an example.

In its present implementation, all IQON variables are alike. Any input from other variables simply modifies the rate of increase or decrease of a variable. Each has a central 'neutral' position at which its output has no effect. Figure 9.3 shows this schematically.

If variable 'A' is above 'neutral', a positive link from it to variable 'B' drives 'B' up progressively until it reaches the limit of its box. Similarly, a negative link to 'B' drives 'B' progressively down. Thus 'A' determines the rate of change of 'B'. Multiple inputs to a variable are simply averaged,

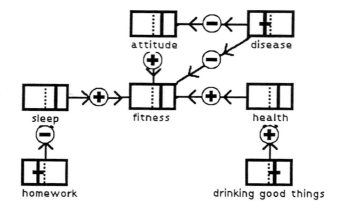

Figure 9.4. *Nancy's IQON model for keeping fit*

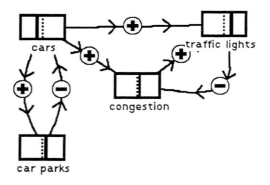

Figure 9.5. *Burgess' IQON model for traffic congestion*

taking account of sign, to determine the rate of change, though some inputs can be given greater weight than others. The response of each variable is made non-linear, through a 'squashing function' which restricts its values to the range minus one to plus one. A variable also has some (adjustable) internal damping.

These features mean that any system of inter-linked variables will have a smooth behaviour, with no tendency for variables to go to infinity or to produce large step function outputs, and that any system will have a unique stable condition from a given starting point.

Figures 9.4 and 9.5 show two examples of models created by pupils aged about 13. Nancy (figure 9.4) sees fitness depending both on general health and on whether one is getting plenty of sleep, and additionally on attitude.

Jokingly, she says that if the school gives her a lot of work to do at home she gets less sleep. Health she sees as affected positively by sensible diet and negatively by disease, in both cases sliding a little away from quantitative variables towards events. Disease has a direct negative effect on fitness, and also an indirect effect via attitude. The point is not whether Nancy is right, but that she has produced a model which is discussable, and whose results when run may surprise her and lead her to think some more.

Burgess (figure 9.5) was modelling traffic congestion. His 'variables' are more like objects than like amounts of something. Because of the feedbacks in the model, when it is run it can give surprising results. Increasing 'car parks' can at first decrease 'congestion' but, because of the loops between 'cars' and 'car parks' and between 'traffic lights' and 'congestion', the model is liable to oscillate. Again, what matters is that this is likely to lead the pupil to reconsider ideas.

Overall, the results of studies with IQON can be stated as follows:

- all pupils could make *some* model;
- half or more made models with fairly sophisticated interconnections;
- those who *made their own* models were more radical in criticizing or reformulating them than were those who were given previously prepared models;
- many had difficulty creating amount-like variables. The tendency was to create *objects* and *events*;
- some pupils could argue about feedback effects;
- most pupils' work produced *discussable ideas* capable of leading to progress in modelling.

In summary, we have a simple graphic modelling facility, for pupils to build such models out of just a few building bricks, and for them to be able to see some of the basic qualitative interactions at work, without yet having to consider exact functional relations between variables. The significant information is in the *qualitative pattern of relationship and change* amongst variables.

In physics, one might *begin* with such qualitative models. Later, it would be time to see how well defined relationships in similar models can give more precise answers, in numerical simulations.

9.3. MODELLING WITH OBJECTS AND EVENTS

If one wants to make computational models with even younger pupils—say 8 to 12 year-olds—then it would seem to be a good idea to model not variables but objects and events. WorldMaker is a system of this kind, designed and written by Dick Boohan, Simon Wright and the author.

A WorldMaker model of sharks preying on fish might look like figure 9.6.

A WorldMaker world consists of objects on a grid. Rules telling the objects what to do are defined graphically. Thus in figure 9.6, the two kinds of object,

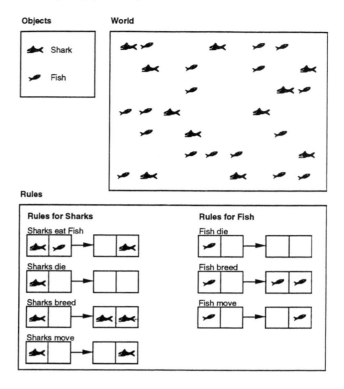

Figure 9.6. *Predator and prey in WorldMaker*

sharks and fish, swim around the grid, being placed on it using drawing tools. Rules are specified by drawings, too. A shark next to a fish eats the fish. A shark on its own may die. A shark next to an empty space may breed or may move. The three rules for fish are similar to the last three rules for sharks. All rules have the form 'condition–effect'. Any rule can be set to 'fire' with a probability selected by a slider bar, so that, for example, relative breeding rates can be altered, or sharks can be made very long-lived. In this model, if sharks breed too fast, they can destroy the fish population and then themselves die out. As is well-known, such predator–prey systems can oscillate.

The concept of WorldMaker derives from von Neumann's idea of the 'cellular automaton' (one of the best known instances being Conway's Game of Life), with the addition of moving objects, each of which retains its identity, and of the possibility of random choices of allowed changes. Each cell of a cell automaton changes state only according to its present state and those of its neighbours. The rules for evolution of the system are local rules, the same everywhere.

A simple model suitable for young pupils addresses the question why

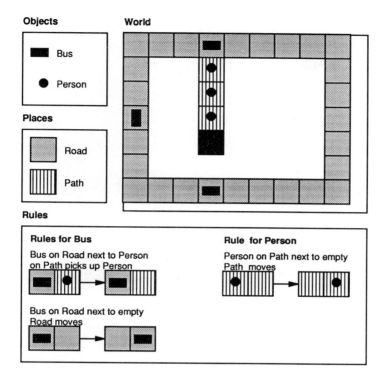

Figure 9.7. *WorldMaker model for buses travelling in groups*

buses in town always seem to come in groups. Figure 9.7 shows the idea.

If buses stop to pick up people, the buses soon become clustered on the road around which they travel. WorldMaker allows directions of movement to be given to an object by the background it is on, making it simple to construct paths or tracks for objects. The example illustrates one of the several ways in which backgrounds and objects can interact, which include changing the other into a different one. An example of such changes is a 'farmer' who moves around the grid 'planting crops' (i.e. changing bare earth to plants) and one or more 'pests' who move around destroying the crops. Another is shown in figure 9.8, in which a creature moves purely at random, but moves more frequently in the 'light' than in the 'dark'. The result is that any initial distribution of creatures ends up with most of them in the 'dark' region.

An even simpler system is able to illustrate molecular diffusion, as in figure 9.9. The walls can be drawn anywhere one likes, and the initial distribution can be varied. The educational lesson here is important. A large scale, macroscopic appearance of systematic change can be generated by what is at the microscopic level random. Exactly the same rule will produce

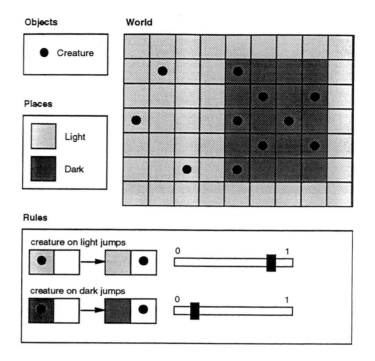

Figure 9.8.   *WorldMaker model of preferential random distribution*

the outward diffusion of particles placed in a cluster at the centre of an otherwise empty screen.

An adaptation of the model in figure 9.9 leads to a model of diffusion-limited aggregation. One just adds another object, a seed, which does not move, and the additional rule that a molecule alongside a seed is captured and turns into a seed. Figure 9.10 illustrates the kind of fractal structure which can result.

Let us mention some other models, simple and more advanced, which WorldMaker makes possible. One is radioactive decay, in which the rule is simply that an object representing a nucleus has a finite probability of changing to a stable nuclide. Such a model is readily extended to a decay chain.

George Marx's book, *Games Nature Plays* gives the example of a forest fire. A cell can be empty, or can contain a tree which is alive or is burnt. Trees are placed at random with a certain density over the screen, and one of them is 'set on fire' (figure 9.11). A tree burns if one or more of its neighbours burns. Will the fire travel all through the forest? It turns out that there is a critical density of trees for this to be likely. An equivalent

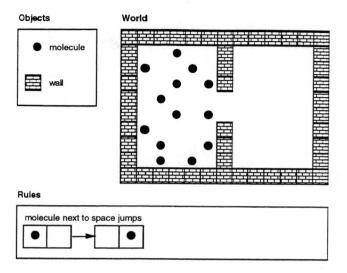

Figure 9.9. *WorldMaker model of molecular diffusion*

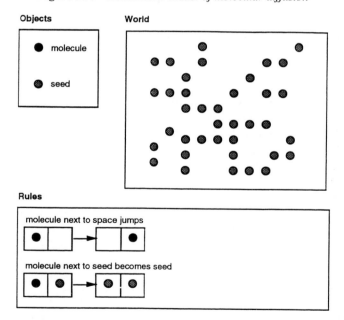

Figure 9.10. *WorldMaker model of diffusion-limited aggregation*

problem is that of whether a mixture of conducting and insulating grains will be conducting, or of whether there are continuous percolation paths for oil through cracked rock strata.

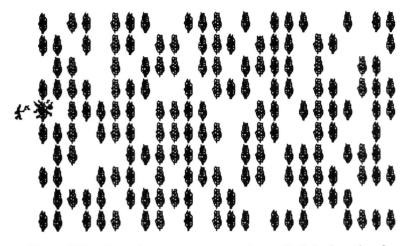

Figure 9.11.   *Forest fire: one tree is set on fire—will all the forest burn?*

Simple examples of chemical reactions can be modelled by having cells filled with two or more species of 'molecule'. Molecules may move to empty cells or may combine with others nearby to make product molecules, which themselves may react in the reverse direction.

All these models have the great advantage that the objects one is talking about are directly represented on the computer screen. If the work concerns sharks eating fish, there are icons of sharks and fish to look at. If the problem is about molecules, one looks at an array of entities representing molecules, not at a display of variables such as temperature and pressure (though the system might in addition calculate these). The behaviour of the whole system is represented to the student by the visible pattern of behaviour of the objects, not as values of system variables. In general, the rules for the behaviour of entities are simple and intuitive, usually relating directly to their behaviour in the real world. Despite this simplicity, quite complex and analytically intractable systems can be studied.

9.4.   REFLECTIONS

The normal order in which people come to appreciate the role of computational models, is far from ideal. One is first supposed to learn functional relations between quantities (Ohm's law, Newton's laws etc.), then some differential calculus, then integration, then numerical methods, and finally one is expected to see the unity in all this. This path is followed hardly any distance by most pupils, and the whole distance by almost none except the best doctoral students.

It may be better to invert the normal order. We should perhaps concentrate from the beginning on form, defined at first loosely and then more precisely. At present we leave form until last, if we ever reach it at all.

If it is true that children would find computational representations of *objects* easier to deal with than representations of *system variables* then this suggests one way of beginning with modelling in which the child tells the objects what to do, not the variables. *Form* is then represented by patterns of behaviour of collections of objects.

A second beginning, directed towards analysing systems into related variables, might be with modelling systems for qualitative reasoning, or patterns of cause and effect, involving variables. Here one has the possibility of looking at form as the typical kind of behaviour of systems with a given structure. The reasons why all oscillators oscillate are fundamentally the same. The reasons why stable systems are stable are often basically similar.

I want to emphasize the very real importance, equally for young pupils and for the best experts, of qualitative reasoning about form. The young child can often guess how things may go, and can look at a model on the computer to see if it 'goes right' or not. The expert is an expert just by virtue of having passed beyond the essential stage of being able to do detailed calculations, having reached the even more essential stage of knowing what kind of calculation to do, and what kind of result it will give.

Finally, we need tools that let children be creative and inventive in their modelling.

When Greek civilization formed... the wisest thinkers brought a new attitude to science: they sought general schemes of explanation that would appeal to the inquiring mind, not simple myths to satisfy public curiosity. Their aim, as they put it, was to 'save the phenomena', or save the appearances, meaning to make a scheme that would account for the facts. This was a grander business than either collecting facts or telling a new tale for each fact. This was an intellectual advance, the beginning of great scientific theory.

Eric Rogers *Physics for the Inquiring Mind* p 223

CHAPTER 10

EARLY ASTRONOMY AND PHYSICS EDUCATION

Maurice Ebison

Philosophy is written in this grand book, the universe, which stands continually open to our gaze. But the book cannot be understood unless one first learns to comprehend the language and read the letters in which it is composed.

Galileo: The Assayer

The Nuffield Physics programme had a profound and beneficial effect on physics education in Britain that extended far beyond those who taught and learnt within the course itself. Eric Rogers, whose influence on the last year of the programme can be seen by comparing the *Teachers' Guide for Year V* with his engaging book *Physics for the Inquiring Mind*, intended that almost one quarter of the final year of the course would be devoted to planetary astronomy used to illustrate the way in which an important theory develops in science. At first sight this emphasis upon the early history of astronomy in a physics course seems rather strange and certainly at the time it roused not a little controversy. Those of us who rejoiced in the treatment of Greek ideas on the structure of the universe felt very much in a minority at that time. It may be useful, therefore, to rehearse the reasons why we applauded Eric Rogers' boldness in first suggesting the inclusion of this material and then his tenacity in rebutting arguments for its removal.

The astronomy was taught not with a main aim of imparting detailed factual knowledge but rather to give students the opportunity to become acquainted with the ingenious paths that men have followed as they have sought not only explanations for phenomena in the sky but also the means to allow accurate predictions of future events. For more than two thousand years these problems were accepted as being amongst the most important that needed to be addressed by scientists. Why do stars seem to form a fixed pattern which appears to rotate round the Pole Star once every twenty-four hours? Why does the Moon also seem to move across the sky east to west

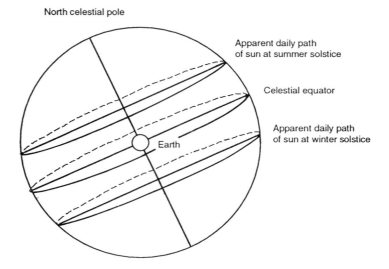

Figure 10.1. *Apparent daily paths of the Sun on the celestial sphere*

in a regular way but with a slow west to east lagging motion relative to the stars that carries it across the sky within the zodiac belt in a month? Why does the Sun seem to move in a similar sort of pattern each day except that its slower west to east lagging motion behind the stars means that it takes one year to complete the ecliptic circle as in figures 10.1 and 10.2? Why do the planets, which in common with the stars, the Moon and the Sun, sweep from east to west in daily motion and like the Moon and the Sun, have a slow lagging motion from west to east along paths in the zodiac belt, also show periodic 'retrograde' motions in which the planet reverses its motion for some time, comes to a stop and then resumes its path once more, as in figure 10.3?

The Greeks asked whether there was some 'actual' motion that could explain the observed motions of the celestial objects. Was there a model of the universe that could be used to explain the various observations and allow the future positions of the objects to be accurately predicted? The unusual motion of the planets was distinctly disturbing to the Greeks since, in their view, bodies moving in the heavens must follow ideal paths, which meant some combination of circular motions since the circle was the only perfect geometrical figure.

In seeking explanations for astronomical observations the Greeks were the first to insist on the distinction between appearance and reality; the first to appreciate clearly how to use models in science; the first to realize the value of employing mathematics as a language of science; and the first to formulate rules of logic as a basis for critical thinking in science. This is a measure of the debt that science owes them.

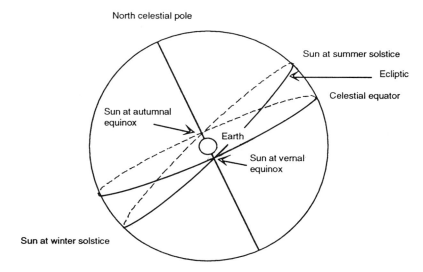

Figure 10.2. *Apparent yearly path of the Sun as seen from the Earth, after stripping out the daily motion*

Figure 10.3. *Part of the apparent path of Mars against the star background, after stripping out the daily motion*

The Greeks produced a number of imaginative and intriguing theories to explain observed astronomical phenomena and Eric Rogers comments: *Those theories were the works of genius...* (Rogers 1967). Two geocentric theories in particular may be used to illustrate this genius. The earliest serious effort to produce a model of the universe was made in the 4th century BC by the great mathematician, Eudoxus. His model, seen in figure 10.4, consisted of concentric transparent spheres with the Earth at the centre. The outermost sphere, which carried all the stars, rotated once every twenty four hours in an east–west direction thus explaining the diurnal motion. It was not possible to represent the unusual and complex motion of the planets using a single sphere, and in the simplest form of the model they required an additional

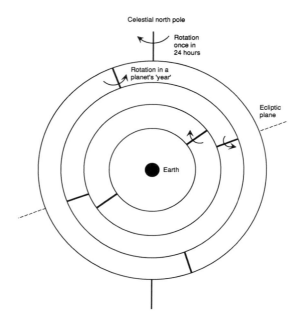

Figure 10.4. *Eudoxus' system of inter-locking concentric spheres*

three spheres. The innermost sphere carried the planet on its equator and its axis was attached to the inner surface of a larger sphere whose axis lay in the plane of the ecliptic and at a small angle to the axis of the first sphere. The motion of these two spheres rotating at the same angular speed but in opposite directions produced the retrograde motion of the planets. The axis of the second sphere was attached to the inner surface of the third sphere whose axis was set at an angle of 23.5° to the axis of the outermost sphere and which rotated in a west–east direction in a period equal to the time taken by the planet to move once completely through the zodiac belt. The five planets known at the time each required four separate spheres, the Sun and Moon required three each, and so together with the one sphere needed by the stars, there were 27 concentric spheres needed to explain the observations. The system was accepted by Aristotle who proceeded to refine it so that it was more closely in accord with increasingly precise observations. He also realized that the complex motion made by one planet's quartet of spheres would be handed on, unwanted, to the next planet's quartet. So he inserted extra spheres to counterbalance the motion between one planet and the next. Ultimately, therefore, he finished up with 55 spheres in a model that began to collapse under the weight of its own complexity.

There were, of course, difficulties in the model of concentric spheres. To take just one example, since each planet remained at a constant distance

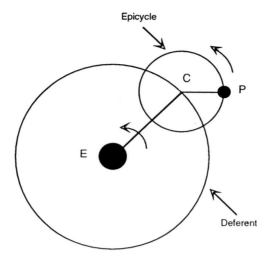

Figure 10.5. *Simple epicyclic model of motion of a planet P*

from the Earth it could not explain why Mercury and Venus in particular appear brighter at certain times than at other times. Nevertheless anyone who examines this scheme cannot help but feel admiration for Eudoxus' skill and originality, particularly when we recall that it was extremely unlikely that he had any kind of physical model with which to experiment. It was a *tour de force* in thinking geometrically. Eric Rogers writes: *In a sense Eudoxus used harmonic analysis—in a three-dimensional form!—two thousand years before Fourier* (Rogers 1960). Certainly the scheme had the merit of using simple geometrical forms (spheres) subject to a simple principle (uniform rates of spin) and was adaptable to fit new data as they became available by introducing new spheres as necessary.

In the second century AD a second great geocentric system, which was to dominate astronomical thought for another fourteen centuries, was put forward by Ptolemy of Alexandria. His book, The Almagest, on the motions of the celestial bodies, is a masterpiece of analysis which used many new kinds of geometrical solution. Perhaps the most fruitful of these was the simple epicycle to account for the retrograde motion of a planet. In figure 10.5 a planet, P, is considered to be moving at a uniform angular speed on the circumference of a small circle (epicycle). The centre of the epicycle, C, also moves at a uniform rate on the circumference of a larger circle, called the deferent, at whose centre is the Earth E.

The idea of the epicycle and deferent is marvelously versatile with several variables that may be adjusted to 'save the phenomena'. For example, we may choose the ratio of the radii of the epicycle and its deferent; we may also select the directions and the rates of angular motions of the centre of

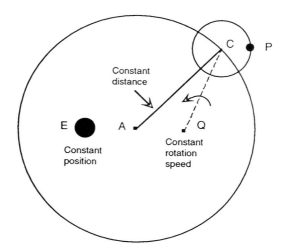

Figure 10.6.   *Ptolemy's more complex epicyclic scheme*

the epicycle on the deferent and of the planet on the epicycle. To illustrate the extreme versatility of the idea it is possible, by a suitable choice of the variables, to produce an elliptical or even a square orbit! (Toulmin and Goodfield 1963).

By the time of Ptolemy, observations of planetary motion were becoming too accurate for a simple epicyclic theory to be adequate. Ptolemy was able to meet the challenge by introducing modifications, shown in figure 10.6, such as offsetting the Earth from the centre of the deferent and moving the centre about which the uniform rotation of the centre of the epicycle, C, was measured to an equal distance on the other side of the deferent centre, A, to a point called the equant Q. All this of course destroyed the beautiful simplicity of the original epicyclic theory, though Ptolemy was able to imitate very accurately the varied motions of all celestial bodies. Ptolemy's geometric analyses were equivalent to a complicated equation for each individual body and the system worked brilliantly by mechanically churning out accurate forecasts of the positions of the bodies for years into the future.

Arthur Koestler in *The Sleepwalkers* (Koestler 1968) takes a rather sour view of Ptolemy's system when he writes: *There is something profoundly distasteful about Ptolemy's universe; it is the work of a pedant with much patience and little originality, doggedly piling orb in orb.* Eric Rogers clearly did not agree with this view, writing: *Here was a gorgeously complicated system of main circles and sub-circles, with different radii, speeds, tilts and different amounts and directions of eccentricity* (Rogers 1960 p 240). I am on Eric Rogers' side in this matter. Perhaps we should leave the last word to Alphonso X of Castile, a great patron of astronomy, who commented when Ptolemy's system was expounded to him: 'If the Lord Almighty had consulted me before embarking upon the Creation, I should have recommended something simpler.'

It is vital that students understand the nature and use of models in science and in this area the Greek cosmological systems have an important part to play. The significance of models and the difficulty of introducing them to students is emphasized by Rom Harré in his *An Introduction to the Logic of the Sciences* when he writes: ... *there are two fundamental ideas which are of great importance and yet which are not easy to bring home to science students. The first, that of a model, is perhaps the most important a scientist needs to grasp. The second is the great complexity of the criteria used in judging the effectiveness of laws and theories in the sciences* (Harré 1963).

The Greek cosmological models provide an opportunity to discuss the difference between asking 'Is the model useful?' as opposed to asking 'Is it true?'. Their construction is also an excellent example of the essential but difficult process of constructing a mathematical model, in a case in which the data (angular positions of celestial bodies as seen from the Earth) are rather well reproduced without the mechanism which generates the motions being very plausible.

The discussion of models leads to a discussion of the use of analogies in physics. When physicists claim that there are analogies between a model X and a physical situation Y, they may be talking about entities of various sorts. The entities under discussion may be *physical objects* as, for example, billiard balls and atoms (Lord Kelvin, in a moment of magnificent frankness, confessed that he really did think of atoms as tiny red billiard balls); they may be *'stuffs'* as, for example, Maxwell's fluid and an electric field; or they may be *phenomena* as, for example, the division of a liquid drop and nuclear fission. The similarities claimed between X and Y may also be of different kinds. X and Y may be similar by virtue of satisfying the same *physical principles*—perfectly elastic billiard balls and gas molecules both operate according to the principles of classical mechanics. A second type of similarity that provides a basis for analogies is similarity in geometrical configuration—a Bohr atom is analogous to a planetary system in that in both cases there are relatively light objects revolving in orbits round a relatively massive central object.

We have been discussing analogies in terms of similarities, sometimes called 'positive analogies'. Clearly there must be some positive analogies between X and Y otherwise the model is useless. But there must also be aspects in which X and Y are different, that is, there must be some 'negative analogies'. It is not productive, for example, to draw an analogy between two red billiard balls. In general we may say that if X were completely different from Y it would be impossible to draw analogies, whereas if X were exactly like Y there would be no point in drawing an analogy (for in this case X would be Y).

Returning to our discussion of astronomical systems, we note that the Ptolemaic model retained its dominant position in astronomy for about 1500 years, for a number of apparently sound and sensible reasons. As a geocentric system it had strong philosophical appeal.

The idea of simplifying the cosmological model by assuming that the Earth spun once per 24 hours and at the same time moved round the Sun once per year in a large orbit had been put forward by Aristarchus in the third century BC. The idea went nowhere for a number of very good reasons. The Earth really did appear to be at the centre of the universe and very solidly at rest. If it spun daily, surely there should be a constant wind in the opposite direction and if it moved in a large orbit around the Sun the fixed stars should surely show some parallax motion, unless they were enormously distant. (The largest such shift is only 1.5 seconds of arc and was first measured only in 1838.) Perhaps equally important for the failure of Aristarchus' system to take root with the Greeks is that he does not appear to have used his system for making predictions of planetary positions. His work seems to have been purely a philosophical speculation, a general picture of how things might be. Although his work had very little influence on Greek thought it is fortunate that his ideas were recorded, since eighteen centuries later they assumed considerable importance by stimulating the thoughts of Copernicus.

Another potent reason for the persistence of the Ptolemaic system was that it was in accord with Thomas Aquinas' widely accepted synthesis of Christian belief and Aristotelian physics. The enduring nature of Aristotelian ways of thinking has to be an important concern for science education and is the second valuable reason for including early astronomical thought in a physics course. Thomas Kuhn in his book *The Copernican Revolution* wrote: *Aristotle was able to express in an abstract and consistent manner many spontaneous perceptions of the universe that had existed for centuries before he gave them a logical verbal rationale... the opinions of children, of the members of primitive tribes, and of many non-western peoples do parallel his with surprising frequency* (Kuhn 1957). The effect of this almost subconscious acceptance of Aristotelian views on physics education is emphasized by E J Dijksterhuis: *To this day every student of elementary physics has to struggle with the same errors and misconceptions which had to be overcome, and on a reduced scale, in the teaching of this branch of knowledge in schools, history repeats itself every year* (Dijksterhuis 1961). Physics educators need to be acutely aware of the fact that students bring to their study of physics a collection of proto-concepts formed from early common sense experiences.

It is important to appreciate that Aristotelian physics had, and still has, a strong psychological attraction. It is an error to assume that the Greeks did not observe the physical world. In fact, within Aristotelian physics, there was a close inter-locking of observations and explanations which covered an immense range of the phenomena that happened to be accessible to unaided observation at that time. In addition, Aristotelian physics was further buttressed by the way in which it fitted neatly into a much wider philosophical system that was an intellectual *tour de force* in its logical coherence. To take Aristotelian ideas about motion as an example, the fact is that our modern view of inertial motion, in which bodies continue in a state of rest or in motion along a straight line unless acted upon by an external force, is not easy to confirm by observation alone. For example, it is occasionally suggested that modern mechanics essentially started when Galileo observed experimentally that two bodies of unequal weight, released from the top of the Leaning Tower of Pisa, struck the ground at exactly the same time. Aside from the point that it is highly unlikely that this particular experiment was ever performed by Galileo (see Cooper 1935), the fact is that in the everyday world a leaf falls more slowly than a boulder. It was natural for Aristotelians to focus on bodies falling through a resistive medium, for this is the kind of motion we commonly observe. Aristotelians would have considered unresisted motion an unrealistic abstraction. Indeed Aristotle himself did explicitly argue that in a vacuum, where the resistance to motion would be zero, all bodies would move with the same speed, and what is more took this as evidence that a vacuum could not exist. Aristotle and his successors did not observe any less than, say, Galileo; rather they viewed things from a different standpoint. This was not because they were less intelligent or less observant, but rather because the more quantitative approach of Galileo was irrelevant to their world-view. To them it was not sufficient that a theory should merely describe and predict the results of observations; it must above all give them a meaning in a wider sense, by showing that the results were consistent with and part of a complete philosophical system.

What was needed at the time of Copernicus was a completely new way of thinking about the world. The same happens in the early stages of everyone's education in physics. Such a transposition is not easy and needs all the assistance that it can possibly be given. Herbert Butterfield writes: ... *of all forms of mental activity, the most difficult to induce even in the minds of the young, who may be presumed not to have lost their flexibility, is the art of handling the same bundle of data as before, but placing them in a different framework...* (Butterfield 1965).

Copernicus began the process of the fundamentally important transformation in the minds of men which moved human beings from a central position in the universe. Previous cosmological systems had been concerned with representing as accurately as possible the *appearance* of the heavens. They did not consider the structure of the cosmos or the mechanism by which it worked. Copernicus was not a great observer and his system was not the result of fresh observations. He relied to a great extent upon Ptolemy's data which themselves were to a considerable extent derived from antiquity. His chief aim was to find a mechanism that would be consistent with the data. For his theory Copernicus returned to Aristarchus' concept of a heliocentric universe, but in relying upon ancient data that were sometimes inaccurate, he allowed himself to be concerned by irregularities in the movement of celestial bodies that did not really exist. This caused him to complicate his system with epicycles and offsets so that it was eventually no less intricate than Ptolemy's at its most complex (Layzer 1984). In its use to represent and forecast celestial phenomena it was only more accurate than Ptolemy's system in some cases; in others it was less precise. The important factor in Copernicus' work, however, is that his system had something that was lacking in the geocentric systems even where they represented data more closely; Copernicus' system had a core of truth in it. By scientific truth we mean not only that 'it saved the phenomena' but also that, in the words of Copernicus himself, it had: 'a wonderful commensurability and... a sure bond of harmony for the movement and magnitude of the orbital circles such as cannot be found in any other way'. Eric Rogers echoes this viewpoint by quoting John Keats at the head of his chapter on Copernicus in *Physics for the Inquiring Mind*:

Beauty is truth, and truth beauty, that is all
Ye know on Earth, and all ye need to know.

Eric Rogers contributed so much to physics education in his own idiosyncratic style. One of the ways in which I personally delighted in his approach was his insistence upon the value of studying the history of science. He loved physics with a passion that was reflected in all his writing and lecturing. We who knew him count ourselves most fortunate.

REFERENCES

Butterfield H 1965 *The Origins of Modern Science* (London: Bell) p 1
Cooper L 1935 *Aristotle, Galileo and the tower of Pisa* (New York: Ithaca)
Dijksterhuis E J 1961 *The Mechanization of the World Picture* (Oxford: The Clarendon Press) p 30
Harré R 1963 *An Introduction to the Logic of the Sciences* (London: Macmillan)
Koestler A 1968 *The Sleepwalkers* (London: Hutchinson) p 69
Kuhn T S 1957 *The Copernican Revolution* (Cambridge, Mass.: Harvard University Press) p 96
Layzer D 1984 *Constructing the Universe* Scientific American Book (New York: W H Freeman) pp 38–42
Rogers E M 1967 *Nuffield Physics Teachers' Guide V* (London: Longmans/Penguin) p 105
Rogers E M 1960 *Physics for the Inquiring Mind* (Princeton, NJ: Princeton University Press) p 230
Toulmin S, Goodfield J 1963 *The Fabric of the Heavens* (Harmondsworth: Penguin) p 154

Most of us have some superstitious beliefs, even behaviours, based on fears or traditions, and held secret. Some of our superstitions are dangerous such as out-of-date medical ideas, and fear of nuclear dangers in power stations.

We need some superstitions and should treasure them if they are harmless. So we should not necessarily try to throw them out as uncivilized. But we might very usefully try to shift some of them to more harmless fields: from medicine and cranky food to... well to what?

For many years I have advocated the teaching of simple planetary astronomy as a move against superstitious astrology. In my old age I welcome an enthusiasm for astrology as a safe refuge from some other more dangerous prejudices. Of course, I myself regard astrology as trivial nonsense but I now suggest a shift to astrology from some forms of medicine and some aspects of nuclear fear. For a longer future, I now suggest we might try moving some public prejudices concerning nuclear fuels onto harmless astrology instead.

Seriously, in dealing with oil resources yet to be discovered I suggest an upper limit: imagine the Earth's surface just a hollow skin completely filled with oil. How long could all that oil last? At the exponentially growing rate of using oil that would last only a few centuries. Unless we can establish controllable nuclear fusion, the near future seems dark and chilly.

Eric Rogers 'Superstition' *Proceedings of GIREP Conference on Energy Alternatives and Risk Education* Balaton, Hungary, 1989

These are probably the last remarks Eric Rogers published.

CHAPTER 11

NUCLEAR LITERACY

Eszter Tóth

11.1. PROLOGUE

—*'The Italian navigator has landed in the New World.'*
—*'How were the natives?'*
—*'Very friendly.'*

This telephone conversation happened 50 years ago, at 3:53 pm, on the 2nd of December 1942, between Chicago and Boston. Enrico Fermi, Leo Szilárd, Eugene Wigner and others had just initiated a self-sustaining nuclear chain-reaction. Einstein said of this moment: *For the first time in history, men use energy that does not come from the Sun.* The door was opened to the use of energy a million times more concentrated than it is in the traditional wood or coal fire. But this also implies millions of times greater responsibility, and responsible decisions require knowledge. In the case of democratic decisions, this knowledge should be present in the mind of every citizen, as a precondition. So, how to teach nuclear physics?

In traditional physics textbooks the nuclear chapter introduces the discovery of radioactivity with some sentences of history, pictures of men with beards, and with the well-known figure in which α goes left, β goes right and γ goes straight forward. The poor students learning these things may very properly ask: *How can I decide for or against nuclear power if all I know is that α goes left or when Rutherford was born?*

Some countries, among them Hungary, try another way: teaching nuclear physics based on the binding energy of nuclei or, better, on the changes of energy of nuclei. In this structure nuclear physics consists of three chapters in the textbook.

I. Experimental discovery of nuclei and the neutron (Rutherford's, Chadwick's and other experiments, shown by computer simulation).

II. Droplet model for heavy nuclei (mapping the energy valley in the plot of energy per nucleon versus the number of neutrons and the number of protons).

III. Applications of the droplet model (radioactivity as 'cooling' of nuclei, fusion, fission, natural and artificial radiations, health effects, reactors, power plants, bombs).

I shall tell you a different story about how to teach nuclear physics. It is a drama in four acts. I was proud when, at Balatongyörök in 1983, Eric Rogers offered to adopt me as his intellectual grand-daughter. I hope he would have liked my drama.

11.2. ACT ONE

Nuclei don't age; their decay is as random as traffic accidents. In 1981, at Lake Balaton during the GIREP Conference on Nuclear Energy–Nuclear Power, Eric Rogers called on the hundred or more participants to stand up, asking them to throw a 1 forint coin every time he clapped. Those whose coins fell with the value facing upwards were to sit down again. Eric counted the people standing after each clap. With this simple game he simulated the randomness of radioactive decay, to introduce the idea of half-life (or as he called it, the halving time).

In the 1980s it was easy to buy vodka, bedsheets and guns from the Soviet soldiers in Hungary. Even Geiger tubes were very cheap. My students built electronic counters to use with the Geiger tubes. From 1982 onwards, more and more Hungarian schools owned simple Geiger counters, many of them able to be connected to a very simple school computer. Once upon a time there was a country (the German Democratic Republic) which produced cheap isotope generators for schools: the generator was just a small plastic pill containing ^{137}Cs. (^{137}Cs is a direct product of fission, one of the cheapest radioactive isotopes.) From this pill one can easily extract ^{137}Ba (the daughter product of ^{137}Cs from β-decay), just by dissolving it out. The ^{137}Ba has a short 2.5 minute half-life by γ-decay. So it was an ideal tool to demonstrate the half-life time during a single lesson period. *Voilà*: the decay followed Eric Rogers' coins law, halving and halving, with some superimposed random fluctuations. Of course, before or after the measurement of the ^{137}Ba activity, the teacher had to measure the radiation background in the classroom. This was a good opportunity to speak about the presence of radiation everywhere in nature.

To understand something in a deep way means not only being able to follow the logical steps of a proof leading to a conclusion. In science one can understand something if one has one's own experiences. Teachers need to offer experiments (and not only simulation and movies) for their students in the classroom, in every subject including the case of nuclear physics.

In Hungary the nuclear section of the syllabus is usually taught in grade 12, generally in March. On a sunny-windy morning in the Spring of 1986 I brought the small Geiger counter and the Sinclair Spectrum computer to the class. The experiment went well. The students were enthusiastic, listening to the clicks. The background radiation on that beautiful morning was 40 counts in a minute. Then the nuclear chapter was over, and students began to prepare themselves for the Matura examination: numerical problems, work with paper and pencil.

11.3.   ACT TWO

In the early morning of Tuesday, 29 April 1986, I found all my students waiting for my arrival at the science lab. They demanded: 'Please, let's measure the background again!' I had not listened to the radio that morning, so I did not understand their reason for asking. The radio had announced the arrival of the 'radioactive cloud' from Chernobyl. It took only two minutes to prepare the equipment, and we measured the background again. The students went crazy when they read 120 counts in a minute!—'Three times higher! Let's open the window! Let more come in!' But when we did so, the activity fell. The activity of the fresh air was only 36 counts per minute.

Today I know the reason: the laboratory had not been used for the past four days. Closed windows, closed doors. Radon gas from uranium-containing bricks used in the classroom walls was able to leak out, and its concentration together with that of its γ- and β-active daughters had climbed higher than normal in the unventilated room. But in those Chernobyl days I forgot all this.

The Hungarian physics teachers can be proud of the smallness of the social impact caused by Chernobyl in Hungary. Many students, alumni and parents came to the schools in those days, not only to ask the opinion of their physics teachers, but also to see whether Geiger counters showed any contamination on lettuce leaves, spinach, soles of shoes, or other things. Both I and many of my students lived on a very healthy diet at that time. We had learnt from our experiments that the fallout from Chernobyl could be washed off the surface of the vegetables. (We had also learnt that mushrooms 'like' both ^{137}Cs and ^{134}Cs, so that the radioactivity of mushrooms was a little bit higher that Summer.)

In those months of 1986 the number of artificial abortions went up by thousands in several European countries. Not so in Hungary! Some months later, nice healthy babies were born to the former students of the Hungarian physics teachers. This is nuclear literacy in the most vital sense possible.

For three years afterwards, measurement of the radioactivity of the soil after Chernobyl was very interesting. But slowly the fallout decayed, and it was not so interesting any longer. Chernobyl was not news any more.

11.4.   ACT THREE

In 1989, I read a short article in the 29 September issue of Newsweek. It was about radon in homes.

I had learned at university in Budapest that, by using a vacuum cleaner, one can collect dust from the air on a layer of medical gauze. And that if there are any radioactive particles in the air, they stick to the dust, so that with the help of a Geiger counter one can detect some extra decay on the gauze. I did not believe that our simple Russian Geiger tube and home made electronics would be able to show any result but nevertheless I tried it together with some of my students. We closed the windows of a laboratory for a day. Next morning, on 5 October, we took a sample from the air of the laboratory using a vacuum cleaner. The background on the clean gauze was 200 counts in 5 minutes. The dirty gauze (after 30 minutes of suction) showed 7200 counts in the first 5 minutes. We measured the sample at five minute intervals for 4 hours. And a beautiful decay curve came out with a forty minute half-life.

But students have inquiring minds. We took a sample on the same day at noon from the open window, from the fresh air. In the first five minutes there were 2100 counts on the gauze. The students were satisfied, I was satisfied: our hypothesis about the uranium-rich brick was verified. But in October the students learnt about statistical physics, so an idea was born in their minds: the radon has a large molecular mass, so according to the barometric formula more radon should be present in the basement of the school than higher up.

Next morning they came early. They took the sample from the basement: 12 800 counts in the first five minutes! Happiness, satisfaction! But luckily there are students every time who doubt. 'Just to check, we should take a sample again from the window'. It was done. And the surprise came: 6900 counts in the first five minutes. Why was it so high in the fresh air? We checked the Geiger counter, and had a huge discussion; in the end one of the students telephoned the Meteorological Institute. They said that no power plant had exploded, and that nothing interesting had happened since the day before, merely that the air pressure had gone down. This was the reason why more radon had come out of the wall and up from the soil. This was enough for my students to propose another new venture.

Dozens of students organized a radon monitoring group in the school. They measured the radon concentration and recorded meteorological data day by day. They even wanted to know what happened at night. Once they organized a four day survey (day and night) taking samples every hour (imagine their poor teacher! 96 hours without sleep). In the Spring they repeated the day and night experiment for 8 days. Fortunately the next year a small group of students invented an automatic system for the day and night measurements. That year was easier for their teacher!

From the two years of these measurements the students learned a lot. They were willing to learn nuclear physics earlier than required by the school curriculum, just because they wanted to join the 'radon group'. They were happy when they discovered that the radon concentration climbed after 11 hours of rapid pressure drops. They discovered the early morning peaks, the role of wind, cloud, fog and smog, and even the influence of weekend traffic pollution.

Later on, some of the Hungarian schools created a network for radon monitoring. We organized students' conferences at which each school reported its work. One of the schools examined building materials, another the soil activity around their school, and another the radon concentration in the water supply. One student group found some correlation between earthquakes and the radon peaks in their school basement. A country school discovered a jump in radon concentration upwards after technicians used bottled gas for welding. In the schools the results of that morning were pinned on the walls of the school corridor. The students spoke to one another about radiation, they were familiar with radioactivity in their school, and sometimes in their homes. They had learned to live with natural radioactivity by experiencing it every day. This was also a kind of nuclear literacy. And it was not in an emergency, it did not require bombs or a Chernobyl. It was just Nature around us.

11.5.   ACT FOUR

At the end of the work on nuclear physics in Hungarian secondary school physics, we give a questionnaire to the students to estimate their own radiation dose. Usually my students' total dose is about 1.5 to 2.0 mSv per year, not including medical X-rays. The average extra dose in Hungary in 1986/87 because of Chernobyl was less than 1 mSv.

Calculate your radiation dose per year!

Cosmic radiation	0.3 mSv
add 0.015 mSv for each 100 m above sea level	....mSv
If your home is made of	
wood: add 0.35 mSv	
concrete: add 0.5 mSv	
brick: add 0.75 mSv	
stone: add 0.7 mSv	....mSv
From the ground	0.50 mSv
From food and drink	0.25 mSv
From the air	0.20 mSv
Air flights: 0.004 mSv per 100 km	....mSv
Colour TV: 0.02 mSv for each hour per day	....mSv

Medical X-rays
  arm or leg: 3.00 mSv
  lung: 1.00 mSv
  stomach: 3.00 mSv
  dental: 0.20 mSv   ....mSv
Nuclear plants
1 hour per day at the plant: 0.002 mSv
1 hour per day 2km from the plant: 0.0002 mSv
1 hour per day 10 km from the plant: 0.00002 mSv   ....mSv
                                         TOTAL: ....mSv

From the dose sheet you can see that two thirds of the natural radiation is generated by radon and its daughters. In creating this sheet I used data given by other countries, since in Hungary we had very poor statistics on radon. The so-called average Hungarian radon activity concentration had been calculated from surveying 122 houses, and was given as 55 $Bq\,m^{-3}$. (1 $Bq\,m^{-3}$ is 1 decay per second in a cubic metre.) It is the α-radiation of radon but more importantly the α-, β- and γ-radiation of its daughter products which affect our biochemistry. Most printed sources state that living for a year in a place having an activity of 40 $Bq\,m^{-3}$ means an average dose of 1 mSv. Some people say that in the open air, 1 metre above ground level, the world average radon activity concentration is 10 $Bq\,m^{-3}$.

In March 1992 I got a 'phone call asking me to go to Mátraderecske, a small village of 2600 inhabitants in north Hungary, in a beautiful valley of the Mátra hills. I was told that there were to be found there cellars with an activity of 200 000 $Bq\,m^{-3}$, homes with 20 000 $Bq\,m^{-3}$, and that even in the open air the radon activity concentration could be higher than 100 $Bq\,m^{-3}$. The whole phenomenon was discovered in February 1992. Because my previous interest in this field was known, I was invited to Mátraderecske to do further studies.

The day I arrived in the village, the mayor of the village introduced the situation, and I realized that perhaps there were more professional white-coated researchers in the village than citizens. I did not believe they needed my help. But having got there, I walked around the streets. I just walked, and talked with the local people about everything. Slowly it became clear that they already disliked the researchers. The researchers just came in when they wanted to, no matter whether the time was inappropriate; they never said anything about their results; no one knew what they had found, whether it was dangerous or not; nor did they explain what this gas which cannot be seen and cannot be smelled actually is.

After this walk around I offered my help to the village on two conditions: I would work together with the local physics teacher, and I would work together with my students. The conditions were agreed: not happily, but they were accepted.

Between then and now, my students have completed a huge 10-month investigation. We measured the radon activity concentration in 433 homes in the Summer period and in 463 homes in the Autumn period, using alpha-track detectors. To check our results we have done hundreds of measurements with active charcoal. Anyone who works in the field as a researcher can understand the volume of work involved. Together with the students we discovered new geological fault lines, and we found a correlation of the acidity of wells and the radon concentration in homes. The students had learnt as much about nuclear physics, radiation health effects and radon as they could, so that they could explain to the local kids and to their parents how to get rid of radon and how to live with it by regulating it through ventilation. My students are aware that if the radon comes only from last February the possible health effects on the population will manifest themselves 10–20 years later. This is the latency time of possible cancers caused by radiation. When my students work in the village hand-in-hand with the local students they know this. But they also know that to create a panic does not help.

The allowed level of radon activity concentration is something between 150 and 400 $Bq\,m^{-3}$. In Hungary there has been no radon law up to the present time. This does not mean that the government and the local authorities are neglecting the problem. They made a very big survey of a small part of the village, and even reconstructed two of the most radon-contaminated houses. The reconstruction cost more than the total value of the rural abode. But when the government money was finished, the research institutes withdrew from the village. Since last Summer only our students have continued the work, now as a human duty.

This January, when we evaluated the Autumn period track detectors it turned out that 25% of the houses were above 400 $Bq\,m^{-3}$, and 28 houses were higher than 800 $Bq\,m^{-3}$. The local authorities and the research institutes obliged us to give our results to them for checking. We have done so. But in the meantime we are not waiting for their tests. We teach the local people how to ventilate their bedrooms. Also we have successfully begun to decontaminate one of the houses. We made long holes through the basement to ventilate one room of the house. In the last four weeks the radon concentration in this room has dropped to one fourth of the original 2000 $Bq\,m^{-3}$, while the neighbouring unventilated room remains at the same high level as before.

My students and I got a much greater prize for our hard work in this village than any adult's respect or money could offer:

My students have learnt in Mátraderecske that to know physics means having responsibility for others. To learn nuclear physics means having responsibility for other peoples' futures.

PART 2

ERIC ROGERS AND PHYSICS EDUCATION

CHAPTER 12

EXAMINATIONS: POWERFUL AGENTS FOR GOOD OR ILL IN TEACHING

Eric Rogers

OERSTED MEDAL ADDRESS 1969

AMERICAN JOURNAL OF PHYSICS   VOLUME 37, NUMBER 10   OCTOBER 1969

Examinations: Powerful Agents for Good or Ill in Teaching

ERIC M. ROGERS
Palmer Physical Laboratory, Princeton University, Princeton, New Jersey 08540
Response of the Oersted Medalist for 1969 to the American Association of Physics Teachers, 4 February 1969.

The pleasure and other emotions that this award brings me are beyond my power of words. I can only say that I am deeply grateful to you and to all concerned for the award of the Hans Christian Oersted Medal for 1969. I take it as a tremendous honor. It relates to my love of teaching, and I want to continue to share that

delight with you—also some new doubts. I would like to confer with you about Examinations as a great power, for good or ill, in our teaching.

Before that, I will show you two things that illustrate my concern with people's understanding of physics. The first is a picture by Goya; *The Sleep of Reason* (or *Reason Asleep*) *Produces Monsters.* There is a plea for the value of our science in lessening fear and superstition. The second thing I will show is an experiment. (Remember the lady who whispered to her neighbor at a serious play starring Paul Robeson, "I am sure they will arrange to have him sing Old Man River somewhere in the play." I heard one of my friends say to another today "I'm sure he'll show a demonstration." So here it is.) A small ring of thin steel, glass-hardened.

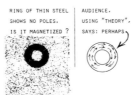

RING OF THIN STEEL | AUDIENCE, | TEST
SHOWS NO POLES. | USING "THEORY", |
IS IT MAGNETIZED ? | SAYS: PERHAPS |

After giving students a very simple 19th Century theory of ferromagnetism, we ask some questions to show the theory bearing fruit, such as, "Is there a limit to the strength of the steel magnet?" and finally, "I believe I have magnetized this ring, yet it shows no poles, no magnetic field nearby.

137

Is it possible that, in any reasonable sense of the term, it *is* magnetized?" Without a theory, the answer is clear: "No poles, no magnet." But with a simple theory many a student makes a good suggestion of a cyclic pattern, and goes on to suggest a test by breaking the ring. We give each student such a ring[1] and when he snaps it he finds a pleasing result. The aim of this experiment is not to show a fact of magnetism, but to illustrate one of the finest values of theory: to provide *new language* with which to discuss and transmit our scientific knowledge.

With those two examples to show my view of our work, I return to my proper theme.

In accepting the Oersted Medal, I feel like a bishop receiving a new and special robe or hat—because you, my lords archbishop, and I, and everyone here, mediate like bishops between physics and people. Here is a picture of our work.

You can see he is a physicist by his big head.

He is conveying physics to future physicists,

engineers, nurses, businessmen, senators .... There is a matter of special feedback for senators

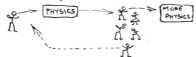

---

[1] These small rings are steel spring washers. They are slightly "dished" so that they fit on a very obtuse cone, but that does not matter. We buy them for a few cents apiece; we make them glass-hard by heating and then plunging them in oil. To magnetize them, we string many on a thick copper wire and borrow someone else's car battery.

which we should label clearly.

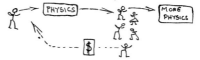

And, in general, future parents of a further generation of students—

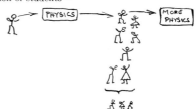

what will they tell their children about science as the children start for school?

We are like bishops—mediating, perhaps preaching too much, but doing our best. I would like to tell you a story, told me by a real bishop recently.

Two devout bishops died about the same time and found themselves approaching the gates of Heaven. They were sorry to find there was a long queue waiting to be admitted by St. Peter, but they stood patiently in line. They turned and found just behind them a girl in black leather costume and a motorcyclist's crash helmet. They were a little uncomfortable—"Why is *she* here?" Presently St. Peter sent a messenger down the line to bring the girl up ahead, and the bishops were really annoyed. When they in turn reached St. Peter they took up the matter with him. "Why should that girl have priority over us?" St. Peter replied, "My brothers, that girl has put the fear of God into more people than both of you combined."

In our work, too, there is an agent of terrific power like that girl, which we dislike, despise, or at least try to discount: examinations. I would like to talk with you today about the great power of examinations. Some wise people say, "If you have a new scheme of teaching, show me your examinations and I can judge it." I would go further: "Let me listen to your examiners discussing and marking, and I shall know the real value and promise of your work." So I wish to take a broad look at testing; but I will not waste

956   E R I C   M .   R O G E R S

time on comparisons of techniques, such as
elegant essays vs objective tests—those five-shelf
pieces of cheap unpainted furniture on which we
rest such pathetic trust. I propose, rather, to
look at the *Sociology* of Examinations: their
interaction with teachers, with aims and philosophy of teaching programs, with students, and
with the future children of students.

We think of examinations as simple, trouble-
some interchanges with students.

Glance at some of the uses of Examinations.

SOME USES OF EXAMINATIONS

&ASURE STUDENTS' KNOWLEDGE OF FACTS. PRINCIPLES. DEFINITIONS. EXPERIMENTAL METHODS. ETC

MEASURE STUDENTS' UNDERSTANDING OF THE FIELD STUDIED

SHOW STUDENTS WHAT THEY HAVE LEARNT

SHOW TEACHER WHAT STUDENTS HAVE LEARNT

PROVIDE STUDENTS WITH LANDMARKS IN THEIR STUDY

PROVIDE STUDENTS WITH LANDMARKS IN THEIR STUDY AND CHECKS ON THEIR PROGRESS

MAKE COMPARISONS AMONG STUDENTS. OR AMONG TEACHERS. OR AMONG SCHOOLS

ACT AS PROGNOSTIC TESTS TO DIRECT STUDENTS TO CAREERS

ACT AS DIAGNOSTIC TESTS FOR PLACING STUDENTS IN FAST OR SLOW PROGRAMS

ACT AS INCENTIVE TO ENCOURAGE STUDY

ENCOURAGE STUDY BY PROMOTING COMPETITION AMONG STUDENTS

CERTIFY A NECESSARY LEVEL FOR LATER JOBS

CERTIFY A GENERAL EDUCATIONAL BACKGROUND FOR LATER JOBS

ACT AS TEST OF GENERAL INTELLIGENCE FOR JOBS

AWARD SCHOLARSHIPS. PRIZES. ETC.

There is no need to read all that list; I post it
only as a warning against trying to do too many
different things at once. These many uses are
variables in the examining business, and unless we
separate the variables, or at least think about
separating them, our business will continue to
suffer from confusion and damage.

There are two more aspects of great importance,
well known but seldom mentioned. First: the
effect of an examination program on teachers and
their teaching—

coercive if imposed from outside; guiding if
adopted sensibly. That's how to change a whole
teaching program to new aims and methods—

institute new examinations. It can affect a teacher
strongly.

It can also be the way to wreck a new program—
keep the old exams, or try to correlate students'
progress with success in old exams.

Second: the tremendous effect on students.

Examinations tell them our real aims, at least so
they believe. If we stress clear understanding and
aim at a growing knowledge of physics, we may
completely sabotage our teaching by a final
examination that asks for numbers to be put in
memorized formulas. However loud our sermons,
however intriguing the experiments, students will
judge by that exam—and so will next year's
students who hear about it.

Here is an example at an early stage of school
or college physics, when students are not yet
resigned or inspired.

(a) In 3 secs ...?

From a high window drop an electric light bulb.
How far in 3 seconds? The night before the exam
the wise student, distrusting his teacher's good
teaching and his own experiments, memorized
the formula . . .

(a) In 3 secs ...?
(b) In 10 secs ?

$$s = v_0 t + \tfrac{1}{2} a t^2$$

(Or possibly in some parts of the world he put it
into a local computer memory, with a ball-point
pen on his bare arm.[2])

² When I made that horrid comment at a meeting,
the Dean of a girls' college reproached me afterwards,
pointing out that in summer exams girls wear no sleeves
and the equivalent resort is to writing on their knees.
However, since mini-skirts, some girls have done more
poorly in physics examinations.

He uses that, gets a reasonable answer.

The problem then asks *"How far in 10 seconds?"* The student knows his duty; he plows on with the formula.

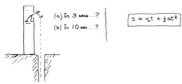

He little cares that his answer requires a hole in the ground, and he does not stop to think about terminal velocity.

Another question: *"State Newton's Laws."* The student is expected to answer in the words of the textbook, even if he does not understand them.

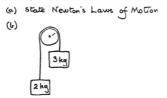

And again a problem for a formula: a wheel and a rope, carrying loads.

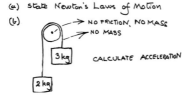

The rope has no mass—a lie; the wheel has no mass, no friction—two more lies. "Find the acceleration." Out comes the formula, from some storage, to "find" *a*.

This is not what Newton gave his Laws for; nor is it worth teaching those Laws for this. As Sam Goldwyn might have said, "if Newton were alive today he would be turning over in his grave."

(I heard one student ask another "What shall I do in the exam?" "Well George, there's always a question with two weights on a string." "What do I do with them?" "You subtract them, you add them, you divide. Then you multiply by *g*." "What's *g*?" "Ah, there you have to spin a coin.

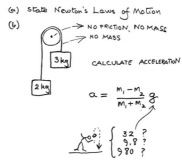

And, George, if you are asked for the tension in the string, don't try it—you would have to think.")

What picture of physics does that leave students with, for later life? If we must ask those questions we should at least ask them in ways that will tell students we expect them to know capably what they are doing; and then we shall also be encouraging ourselves to teach for some understanding. We might change the test to this:

PLEASE WRITE YOUR ANSWERS IN THE SPACES (___) PROVIDED

PROBLEM 1. (a) In the formula $s = v_0 t + \frac{1}{2} a t^2$,

what does v_0 represent?_____

what does $v_0 t$ tell you?_____

the factor $\frac{1}{2}$ arises because_____

Start by *giving* the formula. In general print all formulas on the front of the question paper. We ask, *What is v_0?* and we give two lines on the question paper for the student to write his own reply. Then, *What does $v_0 t$ give?* (If he says it tells us v_0 multiplied by *t*, we agree but pay nothing!) For the answer to *Where does the factor $\frac{1}{2}$ come from?*, we accept calculus, or Galileo's revealing geometry. A sensible examiner, reading a student's answer in his own words, can certainly find out something about that student's understanding. We can ask more:

> (b) A stone is thrown vertically up from the ground, reaches a maximum height of 20 ft. and returns to the ground. If the formula is used to calculate s, will it yield (for the total trip from the ground, up and down to the ground again) 20, 40, or 0 ft.?_____
>
> (c) If the formula is used to calculate the distance a stone falls from rest in 3 seconds, it gives an answer that fits closely with experiment. But if used for the fall in 10 seconds, it fails to agree with experiment.
> Suggest two reasons for that failure._____
>   . . . (5 lines provided for answer) . . . .

For Problem 2 we may ask for Newton's Laws in simpler words. We print the formal wording and ask the student for a colloquial explanation. Try that yourself on a class and you will be surprised at the difference.[3]

> PROBLEM 2.   (PLEASE ANSWER THIS ON A SEPARATE SHEET)
>
> (a) Here are Newton's Laws of Motion. Rewrite them in your own words, to make them clear to a non-scientist.
>
> (LAWS I,II,III PRINTED IN TEXT-BOOK FORM ON EXAM PAPER)

If we must have that sterile Atwood machine, we may *give* the formula and ask for a description of the motion when the rope is massive.

> PROBLEM 2.   (PLEASE ANSWER THIS ON A SEPARATE SHEET)
>
> (a) Here are Newton's Laws of Motion. Rewrite them in your own words, to make them clear to a non-scientist.
>
> (LAWS I,II,III PRINTED IN TEXT-BOOK FORM ON EXAM PAPER)

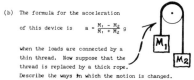

> (b) The formula for the acceleration of this device is $a = \dfrac{M_1 - M_2}{M_1 + M_2} g$
> when the loads are connected by a thin thread. Now suppose that the thread is replaced by a thick rope. Describe the ways in which the motion is changed.

We sketch the device and give the formula, then sketch a variant with a third mass M_3 on a table. Then we ask, "*Where does M_3 go in the formula?*"

> (c) The device is rearranged like this:

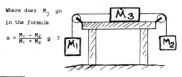

> Where does M_3 go in the formula
> $a = \dfrac{M_1 - M_2}{M_1 + M_2} g$?

---

[3] In one examination experiment, two matched groups were asked to write Newton's Laws. One group was asked to give the formal wording (85% success); the other group, given the formal wording on the examination paper, was asked to explain, in their own words, what the laws mean (35% success).

We can learn much from students' answers, and we can give a suitable grade for understanding. That will take time and trouble from us as teachers, but considering the sociological power of examinations I hold that both making good exams and reading them for understanding are primary duties in our teaching life—to take priority over some lecturing if necessary. (That girl got into Heaven first.)

Critics in one Latin American country told me teachers have too little time, so they must have exams that can be graded by a "computer" (which is what they called the five-answer test-grading machine). I had no time to give them Banesh Hoffmann's telling exposé of the dangers of objective tests. (Perhaps those tests are even turning the more imaginative students away from physics. I am glad to hear in examiners' meetings, "Oh, we can't set that question, it will get 'Baneshed'.") To support my preference for questions with a space on the question paper for the student to write his own answer, I said many of those can be marked by a "semi-computer." A semi-computer is a physicist's wife—an educated nonscientist, programmed to grade short answers by a list of likely forms. Any unorthodox answers, say 10% of the lot, she throws out to a physicist—anyway, she has random access to a physics bank. I am half serious; I would rather do this to preserve the candidate's freedom to give his own chosen reason in his own words.

Anyway, the general difficulty of marking questions for understanding reduces our examinations' *reliability*—reliability in the technical sense of making the marking repeatable from one examiner to another. In some regions, the technical testmaker's craving for reliability is taking attention away from its counterpart *validity*—the test's relevance to aims—and teachers are being urged to test factual bits for the sake of reliability. Testing for understanding *is* often more subjective, but the cost of *less reliable* marking is offset by the *higher validity* of making the examination a physics one, instead of a game in a pentagonal crossword maze. (If you think there is some military overtone in that slur, it is purely coincidental.)

We are all of us clever enough to make examinations that will look for understanding, and will tell students they are looking for that. It is hard work, and needs critical discussion among several physicists. But that discussion does not simply lead to good questions. It also brings out our philosophy—our aims and hopes and the weight we attach to them in teaching—and thus it can control our teaching. In briefing teachers for a new program, I find the disguise of an exam-making conference an excellent starter for a frank discussion of realistic aims.

Yet, framing questions is only half the problem in using examinations to influence teaching and tell students our aims. The marking matters enormously. Suppose we shy away from questions that ask for simple recall of memorized statements, problems that ask for formula-plugging, or, in general, questions that require students to act as rubber stamps. Then we must mark our new questions with an open mind, often accepting a variety of answers.

We can increase emphasis on understanding by the way we assign marks. The way we give credit for separate parts of a question can tell students whether we want scraps of correct items or a genuine understanding. Suppose we manufacture a question with several sub-parts. The usual custom is to give some credit for each part and let a student collect up a passing mark by answering bits of each part correctly. That is dangerous if we are looking for understanding and wish to tell the student so. Some collections of partial answers do reveal general understanding. But in other cases one part of the answer shows the candidate simply does not understand the whole thing. Then that revelation should take charge in marking—it contaminates the whole answer.

To illustrate that, suppose some statistical-minded adviser goes to a pickle factory to rate the quality of the product. He grades the vinegar out of 30 points maximum, the quality of the cucumbers out of 20, the peppers out of 10 and so on, to a total maximum of 100. If the total of his sub-scores is 75, should he say the pickle is 75% perfect? Not necessarily. *One dead mouse in the pickle and the factory is ruined.* I call this the dead-mouse principle of marking. Here is a small tacual example.

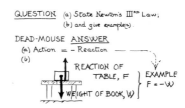

Suppose we offer 5 points for each part. I think we should be as stupid as the student if we just add 5+0 and say he is 50% right.

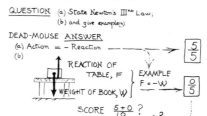

Instead, I suggest we should multiply marks.

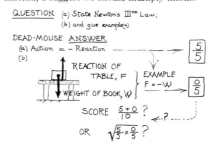

That's the way to enforce the dead-mouse principle. It's also the way to start a protest riot if we reveal it—though the riot doesn't start till next day, because the effect is not obvious. But then we can point out that marks *are* multiplied in an important type of exam which we all meet sooner or later: interviews. Nearly all our students face interviews for jobs, and an interview is an exam. If several abilities are essential in a job, the interviewer tries to test for each; if a candidate fails in any one of those he is refused. Marks are multiplied there—the dead-mouse principle operates.

Some use of dead-mouse-principle-enforcement

960   E R I C   M .   R O G E R S

is not wasteful cruelty; it is important in understanding marking and it is a necessary part of good education for adult life. We can use linear systems that will produce much the same effect; or we can use common sense instead and give a single mark that fits the case, instead of subdividing credits like analytical neurotics.

What do examination marks do to the students who receive them? They can give much-needed judgment, a guiding standard. Most well-balanced students want some tests to enable them to look at their progress. I call examinations used for that purpose "mirror-exams." (A woman can hardly pass a mirror without just a glance to see how she looks; and few of us can resist a chance to listen with an acoustic mirror to a replay of our own voice—however unnerving). The marks for a mirror-exam need not be published with a list of names, that would slant its effect towards vain competition. The results can be placed on a histogram and each student told his place on it. On that histogram we may want to indicate specimen levels of performance, such as "very good," "good," "unsatisfactory"—but those should be based on what we expect of students in this particular course taking this particular test,[4] not on some traditional percentages.

---

[4] In a mirror-exam, failure need not have the insulting import of a final course failure; it may be taken as a frank statement of poor health as judged by this test.

In a large course where I have to assign examination grades ranging from high honors to failure, I try to give the boundaries some objective sense by the following scheme. I ask each member of the faculty teaching in the course to go through the examination questions (before seeing students' answers), imagining a student just at the boundary between passing and failing, and decide how much of each question he would expect *that student* to get right—applying, of course, the dead-mouse principle where suitable. It is surprising how the totals for such a "minimum man" agree from one colleague to another, although their detailed choices differ.

We then make another set of estimates for an imaginary student at the boundary between honors and pass. That "honorable man" seems harder to imagine consistently, so I offer a description. "Suppose a student tutors a neighbor who missed some topics through a few weeks' absence. If he is better than the honorable man he will do the neighbor more good than harm; if poorer, he will do more harm than good—he knows the material well enough to pass, but not well enough to teach."

Then we can use the average marks for "minimum man" and "honorable man" as two fixed points on our scale to interpret grades. We still suffer from subjective judgment, but we do have some flavor of absolute standards.

Then, after a year as mediators between science and people, we end with a final examination. However we construct that examination and mark it, we exert further influences on students by the way we spread the aggregate marks into categories: excellent, good, poor, fail. To the brightest—or perhaps the most industrious—the outcome is an excellent mark, honors, giving pleasure, confidence, and a forward push that many of us have happily experienced in our own student days. To a large middle group we give a lukewarm assent, "you've passed." And to some we say bluntly, "you've failed." We don't do that to any guests after a lunch party. Even if we have taken a dislike to them we don't slam the door on their shoulders as they leave. It seems to me appalling manners to say "you've failed" when a student has spent a year on a course. Worse still, perhaps to fail him at mid-year and then take him back for another half year's failure—invite the guest back to another lunch and slam the door on him again.

I call a final examination that looks back on the finished course an *exit-exam*. Exit-exams are customary, wasteful, and may be rude and damaging to our teaching. On the other hand, an *entrance-exam* to admit to a further course is a different matter. If a student fails that, we say, "I am sorry. You are not prepared." When a student passes an entrance-exam we give him permission to follow the new course, with no further tests except mirror-exams. I am greatly indebted to Professor Leo Nedelsky for this suggestion, which turned up in a discussion of the sociology of examinations.

Let me illustrate that contrast by an example from the teaching of mathematics at school. A school teaches algebra for two years, ending with useless tricks like simultaneous quadratic equations. An exit-exam (final exam) should offer some simultaneous quadratics to be solved. Failure to operate that machinery would lead to a slamming door, "You are no good." What contribution does that make to the country's need for people who have a sense of doing well those things they *can* do, a sense of excellence and interest in their job?

On the other hand, for those who wish to continue into a calculus course, an entrance-exam is fair enough. What would you need for an entrance-exam to calculus? One test to find whether the student has reasonable facility with algebra, binomial expansion perhaps, and mainly, general speed and skill. But a chief difficulty in calculus is the new idea, the concept of a limit. So there should be an important second test: of readiness for new ideas, flexibility of mind, ... And that can be tested, though clumsily. Even if too clumsy, the test itself is a fair warning to students. To those who pass such entrance tests, we offer calculus, to be taken seriously or lightly as they choose, with no further examinations until the next entrance test to a later stage. In other words, the entrance-exam vouches for the course about to be taken, whether it is then done well or not. Again, my thanks to Professor Nedelsky.

Problems of examining lead us to matters of impedance-matching. We enjoy our trade, our own problem solving, design and working of experiments, bright ideas, arguments with colleagues. We can convey those enjoyments to students with similar potential interests, provided we offer them contacts with genuine physics, and not just canned physics of sterile problems or cookbook experiments. Yes, I know, we always say that; but there are some practical conditions for success:

(1) Students must join in the thinking and doing; they must realize that we may have little to do beyond providing equipment and asking questions.

(2) Students need to look for success in work well done, not in a percentage mark. A piece of physics should not be like catching a moving bus—brilliant success for a few long-legged jumpers, painful pass for some with strong arms, supine failure for others. There we can do a lot; we can show the nature of success by our problems and examination questions. We can offer mirror-exams so that a student can see his own progress. We may offer a formal exam question that asks the student to teach us his successful knowledge, or it may be an informal conference—which is an interview, and that is an examination.

(3) We need to match what we offer—topics, treatment, activities—to our students, to make an impedance-match in levels of interest. That impedance-match is easily arranged for very able students, the very bright ones and the ones who already see that they will be physicists. At the high school stage PSSC already does that well, both in suggested teaching and in the suggested

tests. Those tests direct students clearly to thinking and reasoning, and well-understood knowledge.[5] And I have equally great hopes of Harvard Project Physics, and on a wider scale.

Here the impedance-match is being made by the tests and the course selecting students with physics-like abilities and interests. At later stages in college, the good physics student develops his own variable impedance matching to the things offered. And a very powerful agent in changing student's matching is examinations.

But what about the many others, the less able students who do not find our reasoning and discussion and mathematics interesting? There are many in school and college whose type of interest makes a fatal mismatch with physics as we see it. I talk of *interest* because I think that mismatch may often be due to emotional factors (and perhaps cultural traditions) rather than poor mental machinery. We classify students by I.Q. but we hear doubts about the certainty of I.Q. and what it really means. Yet I believe I *know* what I.Q. is: it is a number which indicates probable level of success on I.Q. tests. As such it is not useless, because most high-school examinations and all too many college ones have a strong common element with I.Q. tests—rote memory, use of language (including arithmetic), and simple reasoning. So it is not surprising that high scores for I.Q. predict high scores in academic studies and, consequently, interest in such studies.

For many with lower scores, physics does not seem interesting. Yet we should offer some forms of science in education. In attempting to provide good forms, we run into serious difficulties. We get bright suggestions out of our own ingenuity. Distinguished colleagues press on us teaching that they can do on a one-day visit—but which is not transmissible to teachers who teach 30 weeks a year. (In building a viable project for school physics, I have to censor for "transmissibility".) Or, worse still, we make a vast survey of students' interests and of dissected characteristics of science, and then let the statisticians guide us to an

---

[5] I suggest that the fact that still fewer high school students seem to choose physics may well be a result of PSSC being presented as a genuine sample, which thereby repels those students who would have chosen physics only if it meant formula plugging and short-term memorizing.

organized average in which plain enjoyment and sense of success get lost.

How can we do better for the three quarters of our growing young people whose interests do not match with our formal physics? For them, as with science teaching for younger children, I hope we can avoid producing programs by dilution downward. We take college physics and dilute it so that it is suitable for high school, but I fear further dilution by picking scraps for younger students or slower groups. There is a danger of keeping too much of our own academic attitude, instead of sharing wonder and delight. I think, rather, that the promising programs are those which begin at the bottom, from a careful beginning with real students. We should try to construct treatment relevant to interests and abilities, which is another way of saying "seek good impedance-matching." That's what relevance is—*educational impedance-matching.* I think we should move away from diluted physics, whether scrappy or simplified or ultra clever, and equip a new room labeled SCIENCE AND TECHNOLOGY WORKSHOP.

There we start by not worrying about extracting principles and building theory; we offer things to do. For example, build an amplifier (at the level of doing rather than understanding); experiment with glassblowing, electric circuits, microphone and oscilloscope; make collections and exhibits. Then we need to offer each who enters a better-than-passing grade by encouraging each to succeed along his own line. No competition, no marks, no formal exam, except the most cogent examination of all, personal interviews. With different students this can range from encouraging with real students to praise for a simple collection. Just think of the difference in a science fair between judging an exhibit by the number of flashing lamps and judging by asking the owner to explain how it works. A kindly talk—no inquisition, yet penetrating—not only judges entries on a basis nearer to science but conveys such values to young students. For others, a simple collection may be the medium to provide success.

We seek wonder and delight—intellectual delight for some, and delight in possession for others. And we can distill from our commentary and interviews one common element that we should encourage: a sense of expert knowledge, *a sense of excellence.*

Now, for the first time in history, more than half the working population of a great country is engaged in service jobs, to carry goods or do things for people. Less than half are concerned with producing material goods. How much pride in being an expert, how much sense of excellence, do members of that majority associate with their job? Many see little chance for expert skill, and perhaps some do not even recognize the possibility of a sense of excellence in their own work. I go into stores; again and again I find little pride in expert knowledge of the job. I ask for "some scotch tape that is half transparent; you can write on it with pencil." "Sorry, don't have it." I produce a sample. "Oh, magic scribble tape—here you are." No imagination, no pride in matching wits with the customer. That is a sad commentary on our civilization. The fault lies not with inheritance or heaven, but with school. Education has failed to give experience of a job well done, a pride in thoroughness.

Science can help, though hardly if we grade honors, pass, or fail in an exit-exam. We can contribute if we give praise directed towards the future, so that our tests (if any) take on the aspect of entrance-exams to later life. Then we are encouraging lasting values, giving skill and use of science to some, an understanding of science to some, and to many the satisfaction of personal success in "doing science" in the science and technology workshop.

Thus I hope we can broaden our appeal to a wider spectrum of young people, by arranging a better impedance-match between science and people, and by using examinations both as guides to our teaching and as communication to students—so that our examinations are entrance-examinations to the world, to help each build his own sense of excellence.

Therefore we think of pupils in general. And we think of them, not just when learning physics at school, but a dozen years later when they are out in the world: a young man or woman working in a bank; a lawyer, who must deal with scientists and even with science; a nurse; a shopkeeper; a history teacher in school or university; and the parents of young children who in their turn will approach science with an attitude—of delight or of boredom—that starts at home.

Eric Rogers *Revised Nuffield Physics General Introduction* p 2

CHAPTER 13

ADDRESS TO INTERNATIONAL COMMISSION FOR PHYSICS EDUCATION CONFERENCE
EDINBURGH
29 JULY–6 AUGUST 1975

Eric Rogers

Forty years of physics have not trained me in 'scientific method', which of course should be called 'methods'. The old word 'method' has passed away. It started 350 years ago and died last century. Scientific methods have passed me by in my 40 years of physics... They have not made me tidy, well organized, punctual and unbiased in my opinions. My psychologist friends assure me that it is just not true that we can easily transfer a training in physics, which we so ardently and sometimes angrily impose, to people's lives in general or to other studies. Only if there is a strong feeling of delight, ambition, a wish—a sentiment as the psychologists call it—only then will it transfer. So I have with sorrow heard scientific method being claimed as something we give to the general public. I haven't noticed it, and in 40 years of physics I don't seem to have picked up very much of it myself, outside of physics where I am careful to keep my trade union card...

I come to talk to you, not about the special one per cent of pupils, but about the rest. I shall talk about the general body of physics teaching, but before I do that I would like to show you one or two experiments. Not just as comic relief, as in the extra scene in Shakespeare when it gets too grim, but because I have a message for you concerning some experiments.

My first experiment... is just to remind you that some of the simplest things in physics do not need complex apparatus. I weep when I see manufacturers making elegant devices to throw or release two objects under gravity. Of course the manufacturers are only copying physicists, who are ingenious designers and cannot resist making complex machinery to do what any housewife can do without machinery. First of all I am going to drop two pieces of chalk. I am quite capable—and so is a small child—of releasing two things almost simultaneously. Like that. You do not need a special machine; *it's a crime—don't commit it!* I am now going to release two unequal pieces of chalk and watch them fall. You don't need to hear—just watch, and don't use a machine. There. That's the basis of general relativity. Next I am going to drop another two pieces. This one is going sideways and the other one is falling vertically, simultaneously. I hope they will fall audibly on the bench (I've nearly walked off it—not quite!). That's for the children to try. There is a moral to draw, which is that if you continue to throw pieces of chalk further and faster there will come a time when they go on and on falling at the same rate as the curve of the Earth falls away, and will come round and hit you in the back. Please remember that the Russians did not put up the first satellite. The first one was put up by Sir Isaac Newton, but unfortunately only in a physics book.

My next experiment is to remind you that if you have an object falling for a certain amount of time, and then for twice that time, it falls four times as far—and for three times the time nine times as far, and then for four times the time sixteen times as far: 0,1,4,9,16. Those are the falls. But if you throw it sideways as well it moves equal distances horizontally in equal times, with constant speed. We can illustrate that if we hang stones on a string. This is my suggestion for anyone without any lab, camping in the wilds. You can always tie stones to a string as we have here, and if we have a suitable human aid—you put the youngest child up at the top of the house. Are you ready? Of course there is always the human hope that the operator will fall as well! [The stones, spaced along the string as above, clatter down on to the floor, hitting it at equal time intervals.] Thank you. Don't fall!

Going on from that I have constructed here a mixture of two motions [transparency]. A motion going from a zero line there, then to 1, 4, 9—and going across evenly all the time... you will be bothered by all these criss-cross marks, so we will remove them and have just the parabola, like that, and now I will try to see whether nature will cooperate with me, so that if I provide the initial velocity... nature will provide the necessary acceleration. (*Eric Rogers here throws coins up and along the parabola.*)

I'll ask you one question. Did you enjoy that?... Do you enjoy seeing someone slap someone else's face in the theatre? We do—we enjoy the repetition. We do enjoy even old experiments. Then please let young people enjoy things like that.

I am now going to turn to the general body of physics teaching, not for the one per cent who go on to do physics, nor for the ten per cent who go into other fairly technical things which use physics. This is a picture of the body

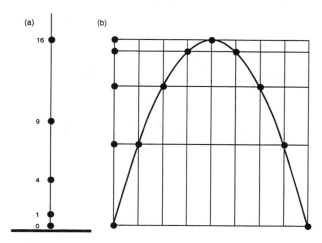

Figure 13.1.   *(a) Falling stones on a string; (b) parabola*

of physics. [transparency] It's a horrible picture. It's nearly all accompanied by lies—hopes if you like, prayers if you will—but lies.

I am not at all sure that physics as we teach it prepares us for life—I suspect that it does the reverse of preparing us for life. I am quite sure it does not teach the love of truth, because that's not what we're doing in physics—we are making models. I am very doubtful about—in the mind of the general public—physics providing for modern technology. It's a nice rumour. I am worried about learning to use scientific methods. And notice my emphasis there—we'd better get into this century and recognize scientific methods as those many things which give us a sense of validity about our knowledge. I am worried about training people to be scientific and accurate, and to think logically.

Up at the top of my picture of the body of physics [transparency] there are some things in a different colour. The patient evidently has some strange green spots on his face. I am the family doctor. I am not the expert on quantum mechanics or dispersion theory... the family doctor is called in. He is much more useful than the specialist. The specialist doctor says, 'Ah, where's my knife? We'll soon get them off!'. The family doctor says, 'Don't worry. His grandfather in Newton's days had the spots, but he recovered. And his father in the last century had the spots, and the uncles have it too. It's in the family—they'll recover.' Therefore, to deal with this patient, which is the body of physics teaching, I say what the family doctor says: 'Take your clothes off—undress'. Now when we undress him [removes overlay] we gain a great deal because we are left with only these suggestions of what the general public—young people—can get from physics. And I suggest that some measure of these are possible. We ourselves enjoy physics. Why not share our enjoyment? That is why I showed the experiments just now, to see if you enjoyed seeing things you know well, but yet can share with me. Let our young—and old—students enjoy sharing physics with us.

Wonder and delight is something we have asked Nuffield teachers about. Did your children... show some wonder and delight? Did they enjoy things? Did they have some emotions which psychologists tell us provide a little chance of transferring to later life, and doing good? If not, our hope is lost. Did they show a little more wonder and delight than in more standard physics, more severely oppressed by examinations, by tests, by learning by heart?—by things which I think are nothing to do with physics—I think they are part of a disciplined system for keeping the kids quiet. The teachers said, 'In the younger years, yes—wonder and delight—but half way through the girls began to worry about the examination, and three-quarters of the way through even the boys did'. I don't believe it. It's the teacher who worried. Intellectual satisfaction is something which probably comes later. Teachers said, 'Yes, we saw intellectual satisfaction, the pleasure of solving a problem, the pleasure of getting the better of an experiment, the pleasure of learning something quite difficult'.

... So in thinking about young people, and the general public learning physics, I am concerned with what we can do to make their learning of physics something that they will keep. They will not keep it if they do not enjoy it. They will not keep it if they learn it by heart—memorize it for regurgitation in an examination.

What can we do for young people? To help our thinking I am going to give you a very strange word: gullibility. Gullibility means in French *aptitude de se faire avoir*; a readiness to be had on. It's a dangerous word; it's neither good nor bad. If we were not gullible we'd never get across the street at the right time with the traffic lights. We'd have no fear. We need the fear in our nerves... Gullibility is concerned with fear; with suggestibility. The confidence man works on it; some priests work on it; some medical advertising works on it; some school teachers work on it. It's a readiness to be deceived, a readiness to pick up a story. *Superstition*. Please don't think that physics will stop superstitions. Superstition is a whole family of rabbits and when you push one rabbit down one hole it digs its way round and comes out of another hole. You can shift superstition, perhaps off something that is dangerous such as health, on to always having seven buttons on your coat because it's a lucky number. In some ways it helps my air travel. On a long journey by aeroplane I trade on superstition: I ask for row 13 and then no-one sits with me and I can lie down and sleep... Astrology I will say nothing about, but you can draw your own thoughts about astronomy and astrology.

Irrational fears, are they all bad? No they are not. Gullibility is also concerned with magic. I like seeing magic experiments. They are called conjuring tricks for the children and they are called physics for adults, and if we're not careful they'll turn into religion for the lot. The good side of gullibility is that we enjoy a trick. We enjoy using our imagination. Without gullibility, without being able to be deceived a little, many comic plays would lose their fun. Much imagination would be restricted. Don't despair over it, but think about whether physics teaching can shift gullibility. Can it possibly shift gullibility from a dangerous region to safe regions? I think it can because we can help young people to see that nature is reasonable. We cannot do so by dictating that to them in words. But we can shift gullibility from a dangerous region to a harmless region.

What we can do for them, I hope, is to remember 'wonder and delight', to remember 'intellectual satisfaction', and to see if we can engage them in doing experiments. If we don't they won't want to learn and think. The story is that stupid people don't like thinking. That's not a bit true: stupid people make some of the best swindlers in the world, but they're not interested in thinking about physics. I see people who are not at all interested in physics, collecting Coca-Cola bottle caps or sitting on railway stations taking train numbers, and they are quite intelligent about what they are interested in. So thinking is a question of interest, and interest may come from doing. So I am thinking about experiments for children, experiments for adults, experiments for people in college and university. And I plead that though some should have complex apparatus and be very grand, others should be as simple as possible. Since physicists are very ingenious we shall have to have those experiments designed by non-physicists, otherwise they will just get too elaborate.

... We have students of all kinds. What does our physics do to them? It depends whether we encourage ambition, delight, enjoying ideals, loving to think, fascination with theory, interest in experimenting. If they get positive enjoyment they will carry it into later life... but we shan't see much of that. I am going to wait. Here are two students... later here is a man and a woman. These are the parents of the next generation. Our students become parents and in turn send their children to school... What will these parents say to their children as they start school? Will the father say to the boy, 'Don't worry about physics—you have to do it. I hated physics when I did it, but I passed the examination because I learned the formulae?' That's tragic for the science you and I love... Will the mother say to the girl, 'Girls are bad at physics—do biology.'? Or will both say to their children, 'Physics is interesting experimenting and clever thinking. Go to school and enjoy it.'? *Interesting experimenting and clever thinking*: if we're clever enough to simplify our physics, to stop worrying about complex apparatus, to try to kill off government examinations, then we might give, in two generations, a delight in physics which may help make a great contribution to the whole intellectual world of mankind.

Edited version of a lecture given at the 1975 ICPE Conference in Edinburgh.

Transcription: Brenda Jennison

Minutes of a Nuffield team leaders' meeting
Imperial College
May 11–12th 1963

These unofficial minutes, consisting entirely of Eric Rogers' remarks, now annotated, were recorded at the time by one of the team.

It is not usually possible to post the man with the screwdriver with the apparatus.

Physics apparatus should be kept simple.

We have no right to make unhappy those people we cannot make good.

We need to produce good physicists, but not at the expense of boring all those who will not be future physicists.

I would like to call Mr Jones Mr Jones and not the murderer as it may affect the jury.

We can't keep open the question whether heat is the same as mechanical energy, if we use the same unit for both all along.

This reminds me of someone's special brand of tomato sauce.

Eric Rogers did not like apparatus which was 'special'.

I don't think we ought to persuade people to go in an un-engined plane.

This is such a nice kitten. We must be careful to see it does not drown.

Compliment to an elegant experiment.

I can only afford a two-bedroom house.

It was necessary to guard against over-expensive equipment. Cheapness meant that there was a better chance pupils could try experiments themselves.

Shakespeare put a comic scene in the midst of his greatest tragedies.

Some light touches enhance any course.

CHAPTER 14

ERIC ROGERS AND THE NUFFIELD PHYSICS PROJECT

John L Lewis

14.1. HOW DID IT ALL START?

I was lucky to be born when I was. There was little change in physics teaching in the first half of this century: there was a concentration on formal definition and the numerical manipulation of formulae, and practical work tended to concentrate on the measurement of constants: Searle's bar for the thermal conductivity of a good conductor, Lees' disc for a bad conductor. Exciting changes began in the 1950s.

The advent of nuclear power in the aftermath of the atomic bombs meant that some modern physics had to be brought into the syllabus. The Association of Women Science Teachers together with the Science Masters Association (shortly to be married together as the Association for Science Education) developed a completely new syllabus, *Physics for Grammar Schools*. This was revolutionary at the time as it moved away from the rigid divisions which previously existed; physics being compartmentalized into electricity and magnetism, heat, light, mechanics, sound and properties of matter. New topics like Waves appeared which crossed the barriers. The new syllabus even included some physics since 1895. The latter came under criticism from HM Inspectorate as it suggested dogmatic teaching. To get over the difficulties, the ASE set up the Modern Physics Committee under my chairmanship and the fun began, for it included such people as Ted Wenham, John Osborne, Geoffrey Foxcroft, Roger Stone, Wilfrid Llowarch, Maureen Hurst (see figures 14.1 to 14.7)—and perhaps most significant of all, two HMI, Dick Long and Norman Booth. Although the Inspectorate had their criticism of the ASE's syllabus, they strove as hard as anyone to see how new ideas could be incorporated into the syllabuses.

Figure 14.1. *Ted Wenham and his Worcester circuit board*

Figure 14.2. *John Osborne working with his electromagnetic kit*

I made a visit to the United States to see the work of PSSC, followed by a term in Germany and the USSR. The impressive thing in both the Russian and Ukrainian republics was the organization behind the teaching and the guidance and other assistance given to teachers. A memorable meeting with the trustees of the Nuffield Foundation on the importance of teachers' guides led in due course to the establishment of the Nuffield science teaching projects. Until that time the Foundation had been mainly involved in medical matters and its involvement with education on a large scale now began.

The physics project was established under the leadership of Donald McGill, a Scottish HMI. Eric Rogers always paid tribute to the sound basis for the project which Donald McGill laid. The whole of ASE's Modern

Figure 14.3. *Geoffrey Fox-croft making slow alternating currents*

Figure 14.4. *John Lewis at the start of 'modern physics' in schools*

Figure 14.5. *Wilfrid Llowarch and the e/m Teltron tube*

Physics Committee was incorporated into the project and ten teams were set up around the country to work on different aspects: mechanics in Scotland; atoms, molecules and kinetic theory in Bristol; modern physics in Malvern; electricity and flow in Birmingham; waves in London; energy in Manchester; teams in Newcastle and York. One of the most significant was the Examination team based in Cambridge. A laboratory under John

Figure 14.6. *Maureen Hurst (then Sister St. Joan of Arc) happily cleaning up after a bromine diffusion demonstration*

Figure 14.7. *Bill Ritchie measuring the acceleration of a falling mass*

Osborne was set up in Imperial College through the kindness of Lord Blackett. Eminent physicists were commissioned to write lead papers on a large variety of topics. A powerful steering committee was established under Professor Sir Nevill Mott.

The Director of the Nuffield Foundation, Leslie Farrer-Brown, made two fundamental decisions. Firstly, he selected his own project leaders, wanting to avoid a committee-led style of work. Secondly, the teams were instructed to concentrate on development work and to avoid public statements and debate for a period of two years: in practice this proved invaluable and is a policy from which other development work these days might benefit. It ensured that the material was well thought out and tested before details were revealed.

So the work began, with ideas coming in from all directions. Eric Rogers was visiting England from Princeton University and his book *Physics for the Inquiring Mind* had recently been published and was well known to the team. Donald McGill suggested that he join the team as a consultant (he was alarmed when he learnt the salary of a Princeton professor, but the Foundation supported the appointment) and a series of papers from Eric were circulated to the teams. Then tragedy struck the project: McGill was taken ill and in March 1963, Leslie Farrer-Brown wrote.

The death of Donald McGill on March 22nd was a severe blow both to the Physics enterprise and a deep personal sorrow to all those who had worked with him. His achievements and his abilities had made him a natural choice as a leader of the enterprise, and it is a tragedy that he should have been taken from us before the work was complete. I am confident that under Professor Eric Rogers as Organizer, and of John Lewis as Associate Organizer, we shall be able to continue the enterprise in the spirit in which he initiated it and in the direction he foreshadowed.

14.2.   THE ERIC ROGERS ERA

So the Eric Rogers era began. He preserved the basis which Donald McGill had proposed: McGill began the course with the Bragg quotation 'Concerning the Nature of Things' and that remained, as did many individual topics. But soon Eric Rogers developed what became his 'grand design' whereby the topics studied throughout the five years of the course were brought together into a 'fabric of knowledge' in which a topic studied in one part of the course was seen to have bearing on a topic in another part (figure 14.8).

Eric felt very strongly about the value of this fabric of knowledge—as he usually felt about many things—to such an extent that when, in later years, it was suggested that certain topics might be left out or become optional, he spoke fiercely about the project 'bleeding to death' if anything was left out. In this he was almost certainly wrong as much of the philosophy behind the Nuffield course has been incorporated into other courses without having to adopt his whole fabric. Nonetheless there was much wisdom embodied in that fabric in the way it holds together and justifies the topics which were included.

Eric Rogers was tireless in his efforts in the months after Donald McGill's death. Draft documents were produced, experiments were perfected, apparatus was designed and close collaboration took place with certain manufacturers. Philip Harris, Unilab, Teltron, W B Nicholson, Telequipment and White Electrical were in the forefront, listening to the project and producing what was wanted, rather than what they thought the project ought to want.

It cannot be pretended that these were easy years. Eric Rogers could talk at very great length; certain members of the original team could not take it and opted out. He was also capable of dropping able members of the team because they were advocating, for example, the use of a joule-meter when he did not want such a piece of equipment in the course—and he did not always do so very courteously. He also made considerable demands on the Director and Assistant Director (Tony Becher) of the Nuffield Foundation, at times straining their patience. But on the whole, those remaining with the project appreciated the value of what was being done and respected Eric Rogers for his major contribution.

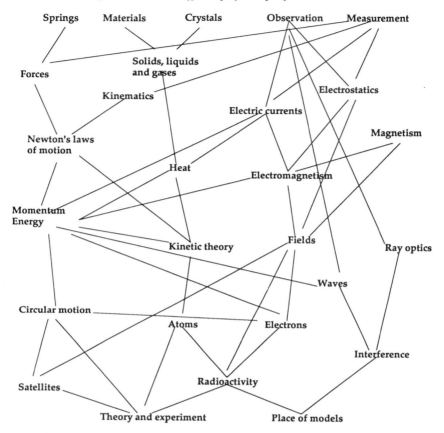

Figure 14.8. *The 'web of knowledge': connections between topics in Eric Rogers' design for Nuffield O-level physics*

For my part, I was based in Malvern and Eric Rogers was living in Hove. We met frequently in London and in the intervening days had long telephone conversations, often lasting for hours. Usually, when I initiated a call, I asked the operator (this was the 1960s!) to advise me of the cost of the call. On one occasion, thinking it would be a short call, I omitted to ask, but it went on for over two and a half hours. I rang the exchange afterwards to see if I could be given the cost only to be told that they were having a sweepstake in the exchange as to the final length and whether it would be cheaper for me to travel to Hove. In fact Eric was right: these calls were certainly expensive, but he wanted to talk to someone to try out his ideas and two hours on the 'phone was a lot cheaper in time and in money than our both travelling to London for a meeting.

The stage was then reached when extensive trials needed to take place so that feedback from them could be used to improve the final published version. Although many of the original team had come from independent schools, it was extremely important that the majority of the trials should be conducted in maintained schools. Eric Rogers set about, with energy and determination, to find the teachers he wanted from different parts of the country. Assignations were arranged, sometimes in strange places, where he would explain what he wanted; Al Parker of Banbury School was 'recruited' on Banbury station and Graham Verow in the depths of Yorkshire. It was not difficult to recruit schools and teachers, not least because the schools were issued with all the necessary equipment. Great enthusiasm was generated, trials were established for the various years of the course in a great variety of schools, and a week-long training course was arranged for the teachers in Loughborough.

The experiment guides were written by members of the team, but the main Teachers' Guides (one for each of the five years of the course) were written exclusively by Eric Rogers himself. He was particularly helped by Hendrina Ellis who typed up all his handwritten material. It was a mammoth task and there were always deadlines to be met. On one memorable occasion Eric Rogers was staying in an Oxford college running a special course during the day and producing material against deadlines all evening. He arranged for Hendrina Ellis to call at 2 am, when he passed sheets of manuscript through a grille for her to take away to type (in those days college gates closed at midnight and no women were allowed inside). This typifies his ways of working and the demands that he did not mind making on other people, and also the devotion of the people working with him. (Incidentally, when it was all over, Eric gave Hendrina two chairs which had belonged to Albert Einstein, his Princeton neighbour and friend.)

The feedback led to modification of the texts and of the experimental work. Other manufacturers began to see a potential market and apparatus had to be approved. Longman was chosen as the publisher, and began to experience how exacting an author Eric Rogers could be. Ultimately the five Teachers' Guides, the five Experiment Books and the five books of Questions were all published; talks were given and courses for teachers were arranged around the country and the first O-level examinations were taken under the auspices of the Oxford and Cambridge Schools Examination Board, eventually reaching almost 30,000 candidates a year.

14.3. EXAMINATIONS

Eric Rogers believed very strongly in the importance of the kind of questions which should be put to pupils. Too often examinations in the past had relied too much on what he called 'cheap recall' and did not promote understanding. He put considerable reliance on Henry Boulind and the Cambridge team, working first on the production of the question books and then on the examination papers. Those papers completely avoided the need to memorize formal definitions: the formulae were always printed on the front of the paper to prevent candidates from thinking that memorizing formulae was what physics was about. Accounts of standard routine experiments were not required. The questions were deliberately designed to test understanding.

Eric Rogers never liked multiple choice questions, then very popular in America. They were avoided for many years; eventually some use was made of them, although one essay-type paper was always included, as he would have wished.

14.4. THE REVISION OF THE PROJECT

Such was the success of the Nuffield O-level physics project that in due course it was necessary to produce a new edition. Teachers felt the need for a pupils' book in addition to the question books. It was therefore decided that the new edition should consist of revised Teachers' Guides which would incorporate the Experiment Guides and a set of Pupils' Books incorporating the Question Books and also providing the text which it was felt pupils needed.

Eric was now a much older man, but he was determined to produce the books himself. In this, tribute must be made to one very remarkable member of the Nuffield team, Ted Wenham. He had led the original Birmingham team, which subsequently transferred to his department in Worcester College of Higher Education and part way through the original project had joined me as a second Associate Organizer. When it came to the revision, Ted Wenham took on the job of assisting Eric Rogers, who was as demanding as ever. It needed someone with Ted Wenham's remarkable patience to see it through. Having in the original project known precisely what he wanted, Eric Rogers was now less sure of himself and liable to change his mind. But eventually Ted Wenham ensured that the volumes came out.

14.5. CONCLUSION

On 22 January 1993, a letter appeared in the Times Educational Supplement under the heading *Please put the fun back into physics*. It began:

Thirty years ago the Nuffield Project ushered in a golden age of physics teaching. Many exciting new topics and techniques were introduced, and the intellectual challenge of the older courses was not abandoned. New courses evolved in a natural way and much progress was made.

There is no doubt that those days were exhilarating ones for those of us who were involved, and the result proved to be great fun for pupils and teachers. That this happened owed much to a great man who led and inspired the project.

How far scientists' theoretical thinking will develop at a given time depends on the state of knowledge and interest—on whether the time is ripe. When the general climate of opinion is ready for a change of outlook or a new idea, a scientific suggestion may take root where it might have starved a century before; and this control of the advance of understanding by the intellectual and social climate is still true today.

Eric Rogers *Physics for the Inquiring Mind* p 347

CHAPTER 15

THE KINETIC THEORY OF GASES

Tim Hickson

Many of us in the early 1960s were uneasy with the type of physics we were expected to teach and with the way it was examined. We sensed that there were better ways of doing things. Heat, light, sound, mechanics, magnetism and electricity seemed sterile divisions when we wanted to explore the interactions and similarities between different branches of the subject. What about the strange and exciting behaviour of the very small and the very large?

My Christmas present in 1964 was a large and intriguing book by an Englishman working at Princeton University. It was Eric Rogers' *Physics for the Inquiring Mind*. When we began to learn about a new O-level Physics course funded by the Nuffield Foundation we were excited and stimulated. What needed to happen to school physics teaching seemed about to happen, and it involved the author of *Physics for the Inquiring Mind*.

Working in Worcester, I went with my colleagues on Wednesday evenings to a course for teachers interested in taking up the new course. It was run by Ted Wenham, aided by Jon Ogborn and Geoffrey Dorling. Nearly thirty years later, I can still remember the sense of anticipation we all felt at the thought of teaching this material. There were lots of good things but, particularly, we were impressed as Jon Ogborn described the work on the kinetic theory of gases. He too was clearly revelling in it.

What was—and still is—so striking about that work?

First, it was an example of the way a scientific theory is developed and of its power to reveal new knowledge: just to set out to teach what was meant by a scientific theory was a refreshing innovation. It showed how work on dynamics could be used to explore the world of molecules and atoms. It gave opportunities to show how simple measurements of room-size quantities can give molecule-size information. It was intrinsically interesting, challenging, different in style from much other work in the course—and fun.

At the beginning of the course, pupils experimented with air pressure, met the idea of the particulate nature of matter, took a first look at Brownian motion and estimated the size of a molecule using Rayleigh's oil film experiment. Later they used a simple kinetic theory in work on the expansion of gases and change of state, suggesting a 'thinking model' of a gas as made up of particles flying about at high speeds in random directions.

At 14+ they began the major work with a review of how we might think of solids, liquids and gases in terms of molecules and atoms. The model was used to help see what happened during the evaporation of a liquid, and how it could produce cooling. All the time, students—and their teachers—were offered thought-provoking questions.

Now students were shown the interlocking relationships between the thinking model, the laws of mechanics, experimental measurements and assumptions that went into it, and the predictions it made. One such prediction was Boyle's Law, which they had already encountered experimentally.

Next, a cunning 'thought experiment' to estimate the speed of a molecule was available for those who would appreciate it. A barometer gave the depth of mercury that would exert a pressure equivalent to that of the Earth's atmosphere. In an example of what Eric Rogers called 'desperate physics', students were invited to imagine that they were living at the bottom of an ocean of air which was as thick all the way up as the air in the room they were in, right up to the top of the atmosphere, above which there was nothing. How deep was this ocean of air? Obviously, higher than the equivalent depth of mercury by the amount that mercury was more dense than air, giving a height of $0.76 \times (13\,600/1.2)$, or about 8600 metres. A rough average speed of a molecule at the bottom of the ocean of air could now be found from the speed it would reach if it fell under gravitational acceleration through 8600 metres, which turns out to be about 400 metres per second. Throughout, the various assumptions made in this piece of 'desperate physics' were discussed with the students.

This order-of-magnitude calculation not only gave a striking result but getting it was enjoyable and an example of what 'real' physicists often have to do. Not all students would be able to see the point of this calculation but those who did—and their teachers—found it stimulating. Rogers reminded those who might consider the assumptions too absurd that Boltzmann had used the same approach to arrive at the Maxwell distribution.

The next step was to show how the theory led to the expression $PV = (1/3)Nmv^2$. It worked because a great deal of thought had gone into the way the argument was presented, including the vital simplifying trick of regimenting random directions of motion in a rectangular box into three 'armies' of molecules, moving to and fro, all with speed v, between the ends, the front and back, and the top and bottom of the box.

It was made clear that the derivation was not to be learnt for any examination and that the result would be printed on the front of any question paper. Thus students were free to enjoy seeing what they had learnt being put to use and could concentrate on seeing where they were going. Where they were going was to use the theory to provide a wealth of predictions.

The formula gave the already known Boyle's Law, as long as v^2 depends only on temperature. To convince students of the value of the theory, they were shown how fruitful it could be in predicting things they did *not* already know. Rearranging the expression $P = (1/3)\rho v^2$ allowed v (about 500 metres per second) to be obtained from experimental values of the pressure and density of air.

Bromine spreading rapidly into an evacuated glass tube gave a demonstration of this high value for air molecules' average speeds. In *Physics for the Inquiring Mind*, Zartman's direct measurement of molecular speeds was described. I and my students used to enjoy an analogue of this in which we sprayed dyed water drops through a slot in a large rotating can. In the O-level course, the change in volume of a liquid turning into a gas showed that the average spacing of air molecules must be about 9 or 10 molecular diameters. Having considered the mechanism of a sound wave in terms of molecules, they could see that sound travels almost as fast as molecules, so that the speed of sound became evidence of molecular speeds.

There was then an opportunity to show gases of different densities diffusing through a porous barrier. Amongst other things, this allowed students to understand uranium separation by diffusion. The Teachers' Guide discussed the equipartition of energy and its relation to quantum theory. This was not to be taught but it was stimulating to the teachers to have this extra information.

Finally—again for those students and their teachers who wanted it—there was an invitation 'using short cuts in a holiday spirit' to see how the theory could provide estimates of the mean free path for molecules, their size and mass, and of the Avogadro number.

Again, this adventure provided students with novel activities. First, the statistics of 'random walk' were explored in a very simple manner. Working on paper with random changes of direction for each step, the class produced data showing that for a walk of N strides the average distance direct from start to finish is $\approx \sqrt{N}$ strides. Then bromine was allowed to diffuse into air and, after a measured time, the rough distance from the place where the bromine started to the 'half brown' part of the tube was measured. Using the mean speed of the bromine molecules, the mean free path (about 10^{-7} m) could be found, together with an estimate for the immense number of collisions experienced by a molecule in that time. It was always fun then to check their previous guesses for the average time taken for a molecule to cross the room!

Knowing the mean free path, students could be shown how to find the 'diameter' of an air molecule (about 3×10^{-10} m). Then the number of molecules in the classroom would be found, and a value for the Avogadro number—more 'desperate physics' but which helped to give that number some substance. Finally, a value for the mass of a single molecule (about 6×10^{-26} kg) could be calculated.

This whole journey was like climbing a mountain: it was to be taken in stages but always with the promise of a sense of satisfaction at completing the climb as well as the joy of the view from the top. At the end, students could begin to see a scientific theory as a useful tool for creative thinking as well as becoming very much more familiar with the molecular nature of the ocean of air at the bottom of which they were living. The combination of something so deeply important with such a creative 'holiday spirit' in achieving it was irresistible.

Since we are asking a question about real nature, only experiments on real nature can answer it... Information from books is useless unless it came originally from real measurements... We might go straight to the laboratory and experiment wildly and boldly, hoping to extract the essential story from a host of measurements. Or we might do some thinking first, guess cunningly at some simple types of motion, calculate the consequences of each, and then go into the laboratory and experiment on the consequences. Both methods have contributed to the growth of science.

Eric Rogers *Physics for the Inquiring Mind* p 13

We wish educated people could know from their own experience that experiment is alert, open-eyed and open-minded putting of questions of nature; a necessary basis for knowledge, but never the whole of our science as we now build it: and that theory is a growing structure of understanding which combines experimental knowledge and imaginative thinking and intelligent reasoning. In short, we want well-educated people to feel that they understand science and the people who practise it, and to know that 'science makes sense'.

Eric Rogers *Revised Nuffield Physics General Introduction* p 59

CHAPTER 16

APPARATUS FOR THE INQUIRING MIND

Jim Jardine

16.1. INNOVATIVE EQUIPMENT

Eric Rogers' recipe was deceptively simple: do, philosophize, understand. Much fun has been poked at the legendary Chinese proverb, '...do and understand', yet Eric Rogers believed passionately in the need for hands-on practical experience. The Nuffield O-level physics project injected into schools a whole new range of apparatus, much of it innovative. This apparatus owed a great deal to ideas drawn from Eric Rogers.

Never before had pupils been able, legitimately, to drench school laboratories as they splashed about with ripple tanks, but at least they could now generate their own water waves and investigate a whole range of wave phenomena including interference (figure 16.1). The properties discovered in this way were shown to be equally applicable to other radiations such as 3-cm microwaves. Simple ray-box experiments confirmed some wave characteristics for light and also demonstrated the principles of real lenses (figure 16.2).

For Eric Rogers, electrostatics was a fascinating and important topic. Modern plastics incorporated in the Malvern electrostatics kit helped teachers to obtain more consistent results than had previously been possible. But many pupils will best remember the spectacular results produced by charging a victim to a high voltage with a van de Graaff generator (figure 16.3). The Perrin tube was then used to determine the nature of the electronic charge, while other properties of the electron were demonstrated with a Maltese Cross tube (figure 16.4).

Familiarity with electric circuits was encouraged early in the course, and the development of the Worcester circuit board facilitated the construction of simple circuits by pupils themselves. A water circuit board analogy for electric circuits (figure 16.5) drew on Eric Rogers' *Physics for the Inquiring Mind* and demonstrated series and parallel connections as well as analogies between pressure difference and voltage, and rate of flow and current.

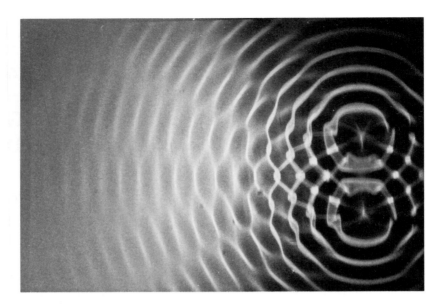

Figure 16.1.   *Interference in a ripple tank*

Figure 16.2.   *Lenses: a view from the author's window*

Figure 16.3.   *Spectacular but safe: the van de Graaff generator*

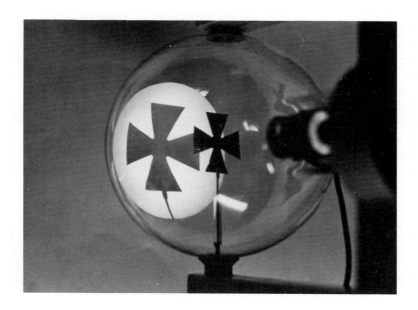

Figure 16.4.   *Electrons cast shadows: the Maltese cross tube*

The structure of the atom formed a major part of the Nuffield course and pupils saw direct evidence of molecular bombardment in Brownian motion. The famous oil drop experiment then gave an order of magnitude estimate of molecular size. It was however with a range of experiments on radioactivity that school physics edged into the twentieth century! The cloud chamber produced startling visual evidence for the existence of energetic particles (figure 16.6), and the Geiger–Müller tube, used with a scaler/counter and some carefully chosen radioactive sources and absorbers, enabled pupils to identify different types of radiations later labelled alpha, beta and gamma. Various analogies, including throwing dice or coins, or a rotating drum full of ball bearings which could escape through holes drilled in the sides, helped to clarify the concept of half-life.

The Westminster electromagnetic kit produced one of the greatest hands-on experiences! With it pupils could conduct all the traditional experiments using small but powerful Ticonal magnets which retained their magnetism despite rough treatment! Forces between magnets and field patterns around them were easily investigated as were the force on a conductor in a magnetic field and the forces between current carrying wires or coils. The catapult forces could be clearly illustrated with iron filing patterns round two current-carrying conductors between Magnadur magnets (figure 16.7). These magnets were also used to build an electric motor (figure 16.8) and a moving-coil meter.

Perhaps the most exciting feature of the Westminster kit was the do-it-yourself transformer. With the laminated C-cores provided, pupils could build their own transformers. Step-up or step-down transformers could be assembled using any number of turns they chose, and a special low-voltage power unit allowed them to find out how to increase the voltage to light a lamp (figure 16.9). These transformers also enabled teachers to demonstrate the operation of high-voltage power lines.

The oscilloscope played a large part in the Nuffield course and could be introduced by such means as a vibrating pen and a roll of paper showing how a sinusoidal curve could be produced (figure 16.10). The oscilloscope and slow alternating current generator demonstrated the leading and lagging effects of different components (figure 16.11), while a large inductor and capacitor produced slow electrical vibrations.

16.2. THE SCOTTISH TEAM

Eric Rogers asked the group in Scotland to develop the mechanics section of the Nuffield programme. Following PSSC we used trolleys as masses and a variety of different models were made before deciding to adopt a 1kg aluminium trolley with a built-in spring plunger. This 'tartan trolley' (figure 16.12) was manufactured locally by John Rollo. By releasing the

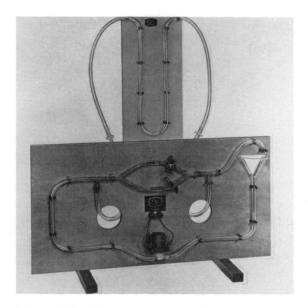

Figure 16.5.   *Water circuit analogy for electric circuits*

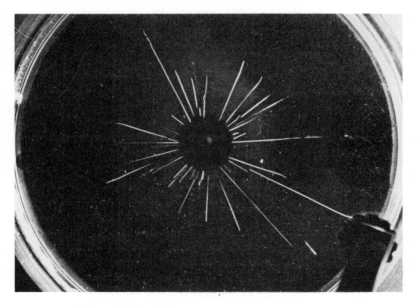

Figure 16.6.   *Alpha particle tracks in a cloud chamber*

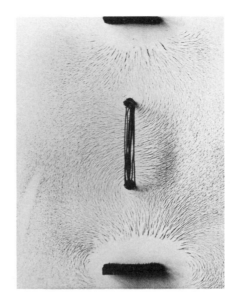

Figure 16.7.   *The 'catapult' field around a coil between two magnets*

Figure 16.8.   *Make your own motor: Westminster electromagnetic kit*

Figure 16.9. *Transforming a voltage: Westminster electromagnetic kit*

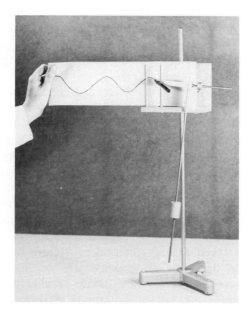

Figure 16.10. *Paper and pen oscilloscope*

plunger pupils could compare the total momentum before and after an 'explosion'.

The application of a constant net force to the trolleys was more difficult. Of course a load (of mass m) could be tied to a string attached to the trolley (of mass M) and led over a pulley wheel. But then the force accelerating the mass M would be $mg[M/M + m]$! An easier way had to be found for introductory experiments on Newton's laws. So we first considered using springs each to be stretched by a fixed amount. Alas, it was impossible to find springs sufficiently sensitive to apply the small forces needed and yet robust enough to withstand pupil participation! Attempts were made to use long coil springs bent over to produce relatively small constant forces but these were voted out by the team. Elastic threads might be the answer. A wide selection of different types of elastic thread was acquired from manufacturers and groups of pupils tried these out with trolleys on friction-compensated slopes. Each thread was attached to one end of the trolley and stretched along its length. Eventually a suitable type was found and eyes were then attached to enable pupils to exert one or more units of force on the trolleys.

Eric Rogers labelled a 'unit' of force defined in this way a 'strang', and his hope was that the Scottish team would be able to concoct a device for measuring strangs—a 'strang-meter'. His idea was to use the threads to calibrate the meter before fixing it to a trolley which could then be pulled along by the previous simple and robust system of a string, load and pulley. The strang-meter would then be reading *directly* the force accelerating the trolley. The load, which might consist of lead shot in a small plastic cup, could easily be varied so that the strang-meter read one, two or three units of force while the trolley was accelerating.

Unfortunately all attempts to construct such a meter resulted in failure. It was never possible to achieve the kind of damping needed to read the force accurately during acceleration. Some twenty years later a French firm produced, and Griffin and George marketed, a dial dynamometer which, when attached to a trolley (figure 16.13), provided exactly the device Eric Rogers had wanted. The meter could be zeroed when the trolley was being pulled at a constant speed by a very small load on the spring. This allowed for friction in the bearings and ensured that the meter read only the net accelerating force. A report on this arrangement appeared in the School Science Review (Jardine 1981).

To measure the speed and acceleration of trolleys the battery-operated 'ticker–timer' from the PSSC programme was developed into a mains-operated, and thus constant frequency, device producing 50 dots a second on paper tape. Pupils could cut up the tape to produce bar charts and see clearly how the speed of a trolley changed under the action of a constant net force.

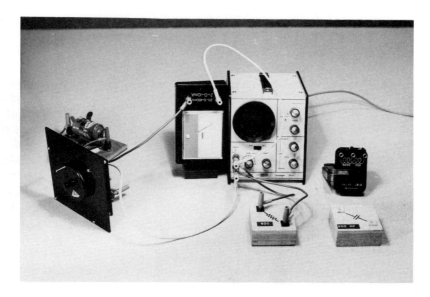

Figure 16.11. *Slow alternating currents*

Figure 16.12. *The original 'tartan' trolley*

Figure 16.13.   *A 'strang-meter'—at last!*

Figure 16.14.   *Constant velocity on a friction-compensated slope*

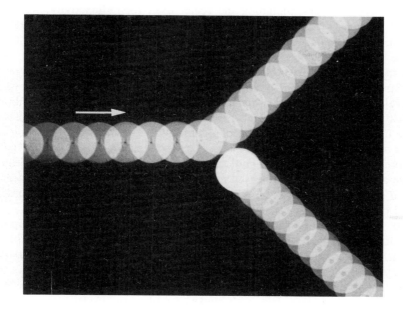

Figure 16.15.   *The 90-degree fork*

Figure 16.16.   *Equality in momentum changes, and in other things*

Electronic timers, including the scaler used for radioactivity, were also used to measure short intervals of time and thus speeds. A photo-electric switch was used to operate an electric clock and thus measure the transit time of a card attached to a vehicle. The speed of a bullet could be found by making it break two circuits a known distance apart.

Stroboscopic photography had been in use for many years but it was the development of Polaroid cameras using very fast film and reliably calibrated multiflash strobe lamps which made this kind of photography suitable for use in schools. It was, for example, possible to show in a matter of minutes that a model car was moving at a constant speed down a slope (figure 16.14). Moreover the actual speed could easily be calculated. Spheres of different size and mass could be released at the same instant and their fall observed. And two similar spheres could be dropped at the same time, one from rest and one moving at a constant speed horizontally.

It was, however, the development of two simple pieces of equipment by John McLaggan, the laboratory technician at George Watson's College, which allowed stroboscopic photography to come into its own. The first was a dry-ice puck made from rejected TV magnets. By gluing a cover on one face and filling the hollow with dry ice the puck could be made to 'float' on a clean glass plate. Two such pucks could be used to demonstrate elastic collisions between two equal masses. If a puck moving at a constant speed collides with a similar puck at rest, the two pucks leave along paths at right angles to one another (figure 16.15). This experiment formed a useful introduction to the interpretation of cloud chamber photographs.

The other piece of equipment he developed was a cheap linear air track made from Dexion Speedframe. It was later produced commercially by Serinco, a firm in Glenrothes. The air track enabled pupils to study constant speed with virtually no net force acting on the vehicle, an elastic collision between a moving and a stationary vehicle of the same mass and the consequences of a moving vehicle colliding with and sticking to a stationary vehicle of the same mass. By using a shutter it was also possible to study the movements of two vehicles before and after a collision.

However, not all the equipment was new or sophisticated. Pupils could pull one another on large trolleys (figure 16.16) as Wilfred Llowarch at the London Institute of Education was showing trainee teachers how to do in the 1950s, and as watched here by the present author, when a shade younger but just as enthusiastic as he is today about providing exciting apparatus for inquiring minds.

REFERENCE

Jardine J 1981 $F = ma$: not another approach! *School Science Review* **63** (222) pp 139–141

Can helpful, enquiring questions really be constructed and used at every stage of our teaching? Yes, but the construction demands considerable time and energy from a group of examiners. If one or two examiners just sit and write attractive sounding questions there is a constant danger of earlier styles and habits taking control. Stronger critical measures are needed. For that, I find that a group of six to twelve teachers, working for a week or more at criticizing and improving each other's questions, can develop an exam which is a strong agent for good. Since in that long session critical attitudes grow and the group can be encouraged to be fearless and tough, I call such a meeting a 'SHREDDER', as a reminder that suggested questions may be treated like paper in a shredding machine!

Eric Rogers *Response to presentation of ICPE Medal* Trieste 1980

CHAPTER 17

'SHREDDERS'

Nahum Joel

How to improve physics education through the construction and discussion of various types of questions, problems and tests

Many generations of physics teachers, all over the world, will remember Eric Rogers for his book *Physics for the Inquiring Mind*, his leadership of projects, or for having had the privilege of being his students or attending his lectures and demonstrations at national or international meetings. All will remember his special gift for transmitting his enthusiasm to others and stirring them into action.

Many fewer had the opportunity to participate in one of Eric Rogers' 'shredding sessions', on which not much seems to have been written. So it seems useful to describe Eric Rogers' approach to these unique events—in particular an international one he conducted in Latin America in 1970— in the hope that this interesting and fruitful technique may be revived and developed further by others.

In 1968, Eric Rogers and I were talking about recent developments in physics education, and were exchanging views on a series of regional workshop-seminars for physics teachers from Latin American countries held from 1965 to 1968 under the sponsorship of UNESCO. Each of these workshop-seminars had brought together about 30 teachers of physics in secondary schools or in teacher training colleges from eight or ten countries, to work under the guidance of leaders of physics curriculum development groups such as PSSC, Nuffield O-level Physics, Harvard Project Physics, and UNESCO's 'Physics of Light' Pilot Project, and other European and Latin American groups.

A number of the Latin American teachers who had come to these regional seminars were using the new ideas in their classroom work and were sharing them with others by organizing local courses and seminars. UNESCO was therefore interested in thinking about the next step forward.

'This is fine', said Eric Rogers, 'By all means carry on. But there is one very important element missing which requires further attention: questions, problems and tests—both for classroom work and for examinations. At this stage, I suggest particularly that we do something about questions and problems in physics to be used by teachers and pupils as part of the process of teaching and learning. Let a group of teachers come to a two-week seminar, each of them with their own physics questions and problems—collected or preferably invented by themselves—and let the discussions take off from there.'

He went on to develop his ideas as to how such a meeting could function. It all seemed so attractive that UNESCO invited him to plan and direct such a workshop-seminar. It was held from the 4th to the 16th of January 1970 in Montevideo (UNESCO 1972). Sixteen physics teachers from secondary schools and teacher training colleges in eight countries of Latin America worked under Eric Rogers' guidance, assisted by Dario Moreno and Claudio Gonzalez from the Universidad de Chile.

The workshop-seminar was remarkable in many different ways. Several of the participants made considerable progress in the art of constructing questions and problems, in recognizing their relevance or irrelevance to what the students were expected to learn, and in seeing how to use these questions and problems as tools for better learning. A striking feature was Eric Rogers' imaginative descriptions and classifications of questions, and his suggestions about how to use each for specific purposes. One particular type of question he called 'water-ski-girl questions' interested participants to the extent that they generated several more examples during the seminar. One of them is reproduced at the end of this note.

It is important to stress that all these various questions and problems did not constitute the main product of the seminar; they were its working medium and the starting point of many fruitful discussions. They led to the most valuable product of the seminar: new thinking.

Typically, a working session would open with Eric Rogers inviting one of the participating teachers to propose a physics question. Other members of the group were then encouraged to discuss or challenge its validity or its usefulness. What is the purpose of this question? Does it really do what you intend it to do? How could this question or problem be rephrased to make it more useful? What learning objectives would it contribute to? How would you use it as part of your teaching? What broader teaching strategy would this imply? Perhaps an experimental investigation on the part of the student? And so on.

The author of the problem would generally first defend it against points raised by the others, but would gradually come to accept criticisms, modifications and further developments of it by the group. It was this tearing apart of the questions that led Eric Rogers to give the name 'shredder' to this sort of workshop.

Exchanges among the teachers were allowed to carry on until they covered a broad range of issues. Quite spontaneously, the group was thus led to discuss matters such as learning objectives and teaching methods—and to formulate their conclusions in realistic terms. The connection between general principles and practical real-life situations was always kept in mind, and attention was paid to the question of what, for given teaching aims and strategy, are the best ways of evaluating the results; and reciprocally, given a set of physics questions and problems, what are the objectives and methods implicit in them. Thus the 'shredder' gave participating teachers practical experience in generating questions to meet a variety of aims, so that if they wished to change their teaching they could construct their own questions and problems to help make the change a success.

The participants reported, both during and at the end of the 'shredder', that they enjoyed the discussions and that, because they were based on concrete situations, they had found them relevant, interesting and useful for their own work as teachers.

It would be impossible to end without a personal touch. For close to thirty years I had the privilege and the happy experience of meeting with Eric in several countries and under various circumstances. I will for ever treasure the memory of those hours spent with him: his engaging personality, his endearing eccentricity, his humane and caring attitude, his inventiveness, his constant bubbling with ideas. After being with him one was always left with the feeling of having learned something exciting, or of seeing something in a new light, and one always felt better.

'WATER-SKI-GIRL' QUESTIONS

A 'water-ski-girl' question gives several answers to a problem, none of which need be completely correct, some of which may contain an element of truth, and others which may be quite false even if temptingly plausible. The student is asked to comment on each of the proposed answers. The name derives from a question in which a variety of reasons were put forward to account for the motion of a girl on water-skis.

Here is an example constructed by the seminar participants:

100 W is equivalent to about 1500 calories/minute. When an electric bulb marked 220 V 100 W is connected to a 220 V supply, it can run for many hours without its filament melting.

Several reasons are suggested below for the filament's not melting. Write a brief comment on each suggested reason. You need not say which explanation is correct, or the best. But you may give your views as to why you consider some to be entirely wrong, and what element of truth there may be in others.

You are expected to comment on each of the statements.

(a) Above a certain temperature of the filament, all the additional energy which is delivered to it escapes in the form of light.

(b) The heat produced by the current is not all spent in raising the temperature of the filament. A good part of it is spent in making the filament and the whole lamp expand.

(c) The filament is made of a metal that has a very high melting point. It would need several weeks of continuous running for the filament to reach that temperature.

(d) As *power = (voltage)²/resistance* and resistance increases rapidly with the increase of temperature, the time comes when the power dissipated by the filament is so small that it cannot warm up any more.

(e) The filament is made of a metal of such large specific heat that its rise of temperature is inappreciable.

REFERENCE

A detailed report of the 1970 Montevideo seminar, written by Eric Rogers, was printed by UNESCO (Paris), reference SC/WS/506, 22 August 1972

Think out your own aims and syllabus. Though yours and mine may differ, neither of us need be wrong... Choose between case histories, demonstration lectures and standard teaching, and historical treatment; and choose your own group of topics. The process of thinking out your aims and planning your scheme will make your course fit you; and it will make you more fully aware of its structure and give it a vitality that is important for a course with such aims. Ask yourself and colleagues, 'Why do we teach science to these students?'... Then go ahead and teach with a happy heart... you and your class can go ahead and think and argue and discuss and learn good science.'

Eric Rogers *Teaching Physics for the Inquiring Mind* p 7

CHAPTER 18

IMAGINE: ERIC ROGERS AND THE NATIONAL CURRICULUM

Bryan R Chapman

This chapter is a flight of fancy about the Eric Rogers of my imagination. It follows from this that my Eric Rogers shares my views on the National Curriculum and all its likely consequences. What relationship these views bear to those the real Eric Rogers might have held I leave to others to judge.

It is the year 2000. The time capsule lands; Eric Rogers steps out. He looks around at the science curriculum landscape facing him. Does he recognize it? Does he see anything in it that makes him believe the time capsule has taken him forward in time? Perhaps. After all, Nuffield apparatus, some of it original, would still be in use. Whether he would find any traces of its underlying philosophy is another matter altogether.

The first thing that would strike him would be that he no longer had freedom to explore the landscape. The boundaries of the National Curriculum Theme Park define both what could, and what must, be explored by visitors. He would find entrance to the park under the control of National Curriculum Daleks who would assure him, in the manner of Henry Ford, that he was free to explore the park any way he liked provided his route took in all the signposted Attainment sites in the order set out by law in the Park's Bye Laws. Would Eric Rogers have submitted to such conditions had the park existed forty years earlier?

Time travellers are resourceful, so let us assume that Eric Rogers manages to get past the Curriculum Daleks. Once in he might even find one or two disaffected Daleks also intent on subversion. But the odds are stacked against them. The first National Curriculum Dalek Master, Baker-Davros, had set up strategically sited checkpoints through which all park visitors must pass. Private security firms of Assessment Daleks police progress through the park and ensure that all visitors traverse it in the order and manner laid down in the Bye Laws. Such policing has clearly had the effect of discouraging all but the most trivial diversion from those rules, and Eric Rogers would find it very difficult indeed to get visitors to explore any site which did not appear on the Assessment Daleks' checklists.

This built-in tendency towards curriculum uniformity would surely depress him. The Nuffield O-level Physics project with which Eric Rogers' name, together with those of John Lewis and Ted Wenham, is so closely associated, introduced diversity into what had become a stagnating curriculum.

Because of this diversity, curriculum evolution was possible. Nuffield was an influence, not an authority. This lesson appears to have been lost on those responsible for the prescriptive design of the Theme Park. The decade of Theme Park policing prior to Eric Rogers' time travel to it will undoubtedly have resulted in a level of curriculum uniformity and stagnation even greater than that which existed forty years earlier when the Nuffield projects were launched. Davros will have won.

Eric Rogers would have made wicked fun of the way the National Curriculum Theme Park is plastered with signs and labels, every part with its prescriptive and often unintelligible name. Whether he would have much sympathy with the emphasis given to practical skills is also open to doubt. Contrary to popular myth, Nuffield O-level Physics was not about practical work or about learning by discovery. It was about ideas. Nuffield Physics was *physics for the inquiring mind*. You do not have to have practical skills to be a good physicist or to appreciate what physics is about; you do have to be able to think critically and imaginatively.

As far as physics is concerned, Eric Rogers was a devoted Newtonian. This devotion did not however lead him to believe that the learning of physics was a mechanical affair. He would have viewed with grave suspicion the behaviourist mould in which the National Curriculum's ten-level Statement of Attainment litany is cast. In the year 2000 he might have turned to the theory of chaos as an image for the complex, wayward and emotionally loaded paths learning takes, to combat the mechanistic image of learning embedded in the Park's Bye Laws.

Printed opposite the title of this chapter is some of the advice Eric Rogers offered about how to plan a physics course. It has a humanity and a respect for the professional integrity of teachers which is totally absent from a system which treats teachers as National Curriculum Theme Park guides.

There are, of course, a few key Attainment sites that no one would wish to leave out of any guided tour of the science landscape, even though the Official Guide Book cloaks them in inexplicable anonymity. The Newtonian peaks are a must on any tour. But does everyone really have to have 'investigated the behaviour of bistable circuits made from two logic gates' and to have 'considered the role of bistables in simple memory circuits to perform useful tasks' before they can be deemed to have made an adequate exploration of the science landscape?

The Nuffield O-level project was set up, at least in part, as an attempt to bring science education into the twentieth century. The National Curriculum Daleks prescribe visits to DNA and to 'current theories of the origin and future of the universe'. But how can young people make any sense of 'current theories about the origin and future of the universe', when they are following a curriculum that does not contain the science from which these theories derive? Contrast the way the Nuffield O-level project paved the way for understanding of large and important theories. Is it really possible to design a twenty-first century Science Curriculum Theme Park which leaves out the key area of twentieth century science? How can one be modern without visiting at least the foothills of quantum theory? Even Nuffield O-level's excursion into wave–particle duality some thirty years ago finds no place in the Park's Official Guide Book. Yet today we take for granted a whole range of artefacts which only exist because of the success of this theory. When new developments take place involving fundamental science, only those who have some knowledge of quantum physics will have any chance of comprehending them.

Returning in the year 2000, Eric Rogers might see traces of alternative routes through the Park which some early national curriculum entrepreneurs had tried to establish when the Park was first opened. However, schools learned very quickly that it was not in their interest to select courses on educational criteria. Their best bet was to select courses on which their students were likely to score best as they passed through the Theme Park checkpoints. By the year 2000, checkpoint performance had become the overriding consideration in the choice of course to follow.

Assessment divorced from aims was always anathema to Eric Rogers. He held dear the principle that one taught a course, not a syllabus. Examinations must be designed for courses, not courses for examinations. Everything about the design and policing of the National Curriculum Theme Park would offend him. And, having not been exposed to the ten-year-long assault on liberal educational values that had preceded its establishment, he would have no difficulty in recognizing the totalitarian intentions of their designers. (Totalitarian—*permits no rival loyalties or parties*, Oxford English Dictionary.) Eric Rogers' own life spanned a period which would have left him with no illusions as to the threat totalitarianism, whether of the right or the left, poses to liberal educational ideals.

By the year 2000, the gap between school science as defined by the boundaries of the National Curriculum Theme Park and the real science going on outside is likely to appear unbridgeable. What has school science got to do with the advances reported each month in *Physics World* or *Scientific American*? Will the journey every young person has to take through the Theme Park really develop the kind of public understanding or appreciation of science which may be necessary if science is to continue to flourish in the next century, especially when this public has been led to expect science to be in the service of the economy? By the year 2000, if not already, it will not be *Physics for the Inquiring Mind* so much as *Physics for the Acquisitive Shareholder* that matters. But what if Acquisitive Shareholders are not convinced about the importance of Physics?

Eric Rogers would have had no easy answers to such questions. Or rather, his answers would still be the timeless ones he always gave. So there is no happy ending to this imagining. Except perhaps to hold to the view that it is the quality of the teachers that really matters, not the curriculum they are obliged to deliver. I am sure this is a view the real Eric Rogers would have shared.

Let us take a fresh honest look at our aims and methods, before the bell rings to start the next lecture. Here I can only give my own aims in teaching physics to general students: my aims now, after many years of aims that seemed good but were justified by wrong reasons. I now want to teach—or rather let students learn—in ways that offer benefits that last a lifetime. That can happen, if our students experience some wonder and delight in what they learn and gain pleasure in intellectual satisfaction. Then they will be proud of understanding—and that can last long.

Eric Rogers *Response to presentation of ICPE Medal* Trieste 1980

When the Nuffield Science Project was started, I became involved as Chairman of the Committee for Physical Science. Although I have never taught in a school (I paid many visits to Colleges of Education and schools) I saw from the outside a vivid picture of their problems. During this time I got to know Eric Rogers well. Looking today at his massive Physics for the Inquiring Mind, with its quotations from Galileo's writings and so much else that one does not find in most physics text books, I am reminded of his vision of what an education in science can be. Our debt to Eric Rogers is very great indeed. The problem of science for all versus science for future scientists was at the centre of our discussions. Even now, as we agonise about Britain's future in Europe, the future physicists will probably drop all study of a foreign language as they start to specialise in the sixth form. In my day, as a young researcher, the language of theoretical physics was German, and from this time some words from German, such as 'Eigenwert' have survived in the language of physics. But now, wherever physics is done, English is spoken and our young scientists come through school and university with a unique ignorance of foreign languages. I don't know what Eric Rogers would have thought about that.

To be involved in the problems of school teaching over a couple of decades was a privilege which I deeply value. And in this a friendship with Eric Rogers, with his concept of what physics teaching could be, was a highlight for me.

Nevill Mott *My involvement with school science*
(Professor Sir Nevill Mott, Nobel Laureate, chaired the Nuffield Sciences Steering Committee).

CHAPTER 19

TRIALS TEACHER

Brenda Jennison

It is no exaggeration to say that Eric Rogers affected a whole generation of physics teachers. His voice can often be recognized in their conversations and writings; as indeed it can be in this book. His passion for teaching physics left a deep impression on those he taught, which for me began in the summer of 1964, the year in which I began my teaching career.

When I was at secondary school, courses in school were defined by examination syllabuses drawn up with the future physics specialist in mind. The content set out to ensure that all entrants to higher education had covered the same material, though the teacher might well not have chosen to follow all of it. Text books, many of which had served physics well for over half a century, were organized around mechanics, heat, light, and electricity and magnetism. As always, the examination papers showed teachers what the real syllabus was all about; 'spotting' questions was an important art, possible because questions had changed little in style for many years. Typical physics questions asked for a definition, followed by an experiment and a calculation. The physics of the twentieth century might never have happened. Some light shone in the pages of journals such as the School Science Review, where enthusiasts described new experiments and demonstrations, but many pupils were fed a dreary diet of dictated notes to be memorized and regurgitated in the examination.

Then PSSC and the Harvard Physics Project came from the USA, and in Britain the Nuffield and Scottish Courses came into being, designed mainly by school teachers with guidance from university physicists. When I graduated, I had no idea what to do next and like many others I drifted into teacher training, happening to choose to do so in Cambridge. My physics tutor would have been Henry Boulind but we were not destined to meet. He was away, busy (I discovered much later) writing the question books and examination questions for the Nuffield Physics Course. I was taught instead by one of the members of the Nuffield Advisory Committee, Clive Williams, who was also one of Her Majesty's Inspectors of Schools. In those far-off days it was thought good that Inspectors should be given time to pursue other interests, which he decided to use training teachers. At an early stage of my training I was aware that something was happening in physics education. However it was confidential, to do with the way in which the project had been set up. As I look back at my notes I realize now that I was hearing first hand what was happening to physics education in England. Pieces of borrowed equipment, three-centimetre waves, ripple tanks and circuit boards floated before my eyes. I certainly had not seen these at school myself.

My first teaching post just happened to be in one of the fifty schools which was to try out the new Nuffield materials, although again I knew nothing of this when I applied. Shortly after my appointment I met Eric Rogers for the first time. He was touring the country meeting all those teachers who were to try out the new course. I met him in my tutor's office. As he did with many other trials teachers, he came to tell me about the course. As I opened the door, a bit apprehensive about meeting the author of *Physics for the Inquiring Mind* and still a student with a promise of a post but no experience, out shot a hand and a voice put me instantly at ease, saying, 'You must be bigger than me'. This was one of Eric Rogers' kindly ways— he could make people feel at ease. He explained that he was writing a new course in physics which would not reward rote learning but rather would encourage understanding. If pupils understood what they were doing then they would not need to memorize for examinations. The questions would be designed so that they could think out the answers from first principles. It was a short meeting. It had to be as I met him in the lunch hour between two examination papers, the authors of which knew nothing of this new philosophy! I remember nothing of the examination papers that I took that day, but I remember that interview clearly.

All the trials teachers met in Loughborough for a week that summer and what happened there set in motion many changes in physics education, aspects of which are still in evidence today. A quarter of a century later when the National Curriculum was introduced, the contrast could not have been starker. Before I arrived at the conference I had met one of the authors and, even before I started my job, I was to be shown how to teach the new course.

The first thing that struck me I found hard to understand. It was Eric Rogers' insistence that the course could not be written as a syllabus. Even the examination board had accepted the draft Teachers' Guides, Experiment Guides and Question Books as their working documents. A syllabus at that time was rarely more than a page long and examination boards made much of the fact that they did not tell teachers how to teach. This course was to be a total package consisting of the content, the apparatus, a philosophy and an examination, all of them in tune. Its philosophy was crudely summed up by the proverb, 'Hear and forget, see and remember, do and understand'. Eric Rogers spoke of the 'wonder and delight' experienced as theories and models were developed and the 'wide-eyed surprise' that pupils should experience in playing with a home-made telescope. He was a master of the art of the simple demonstration. One day we found two bricks waiting for everyone. These were needed to feel 'the spring of the earth', in learning about gravity. I am afraid that I did not appreciate the physics of it at the time but I have used it as a demonstration ever since. By the end of the conference I had been taught about the course I was to teach to my 11-year-old pupils in the coming year, I had tried out the experiments, I had discussed some of the finer points of physics and I felt pretty confident that I could cope.

It was at the briefing conference that I was first introduced to 'shredders'. These are not for the faint-hearted. Participants write questions and copies of all the questions are circulated to the whole group, and each question is discussed in turn, after which participants can submit an improved version. Their real purpose is not so much to write questions, but to think through the philosophy of the course. Throughout the lifetime of the Nuffield O-level Physics Examination all the examination papers went through this process and I learnt a lot of physics.

Eric Rogers encouraged pupils to take home equipment and to show their parents what they had been doing. He strongly believed that teaching father or Uncle George would have a lasting benefit. His practical philosophy was that, if you trusted children to take equipment home, then it would not be stolen. He endowed a fund to pay for anything which was lost. It was rarely drawn upon. He encouraged pupils to be scientists for the day. He knew them well.

Eric Rogers had also encouraged us to abandon the usual 'method, results and conclusion' format for writing about experiments, suggesting asking for a diary in the style of a research worker's note book. He knew how important it was for pupils to write down their own understanding in their own words. Later, when I did it, I was lucky that my Head of Science agreed with all this, because my pupils' books did not look as neat as they used to. A geography teacher complained bitterly that their map-drawing skills had suffered, because we no longer insisted on using a fine mapping pen to label diagrams!

His approach to designing any course was to ask, 'Who is it for?' and, 'What is it that they should understand about physics?' He taught us to design a curriculum backwards, deciding where you want to go, then designing a course to get there. The whole course was to show an interconnected web of knowledge and how a few ideas come together in the great theories of physics. Not for him the Davy lamp just because it is an old favourite.

His end points were few: the kinetic theory, models of the atom, particle–wave duality and the grand theory of Newtonian astronomy. To get there, conservation of energy, electrical circuits, mechanics and waves were needed, each being part of a connected web of knowledge. Ideas would come round several times, simply and concretely at first, and with time for pupils to enjoy playing with ideas and equipment. Later they would revisit earlier ideas and look harder at the evidence for them. He knew well how children's thinking develops, and it is no surprise that the Nuffield course came well out of the analysis of those who looked at it through Piagetian spectacles.

The course was really about the great ideas in classical physics with a hint at the end that there might be more to the subject. Pupils were not given text books but rather question books which would lead them through the ideas rather than telling them the answers. The teachers' book also omitted the answers so there was nothing for it but to work them out oneself. Eric Rogers was a great story teller and he told stories about physics. He trained teachers by telling stories and over the years I have listened to him and, as the story drew to a close, I always became more worried because I knew that I was being tested. He would ask a simple question and I would have a simple answer, but dare I give it or had I missed a vital clue?

After the briefing conference, trials began. The school had already been equipped the term before, with help from a team member in ordering what was needed. I had my teachers' Guides and Experiment Guides and felt ready to go, with a whole set of lesson plans ready made. How naïve I was! We met at weekends to discuss what we had taught, what we were about to teach and to receive any new information. We wrote comments on spare copies of the guides and sent them back. Help was no more than a phone call away. I learned what to do not from reading books or curriculum guides, but by talking to other people, for example, by going to other schools to learn how to use new equipment. Towards the end of the year, courses were held to introduce the second round of trials teachers to the materials. I well remember one experienced teacher saying, 'The theme of the first year is the concept of the atom'. Was it really? I realized that I had taught all the topics—crystals, elasticity, pressure—but that I had not seen beyond them. It had quite passed me by that I should have been taking every opportunity to stress that what we were doing was finding the evidence which supported the atomic model.

Later, I was drafted into the examining team. All the questions for the Nuffield O-level physics examination went through the 'shredding' process. This was when I really began to learn physics from Henry Boulind, John Lewis and Ted Wenham. I remember the day one of my questions actually got into a paper. Ted Wenham said to me, 'You have stopped trying to impress us with your questions and you are now trying to impress candidates'. The way questions were marked had Eric Rogers' stamp on it. He developed the bonus mark system for answers which were better than expected from a pupil at that stage. He encouraged marking in 'piles': 1 for getting started, 3 for a reasonable answer and 5 for mainly correct, which left marking to the professional judgment of the teacher, avoiding rigid mark schemes and accepting variety in how pupils answered. He introduced us to the 'dead-mouse' principle: no marks for answers which looked all right until some remark showed a total lack of understanding.

Here are some of the main ideas I learned through all these experiences.

- Science lessons can and should integrate theory and practical work. Pupils can be 'scientists for the day' as they carry out their experimental and investigational work.
- Good science can be taught to young children, even in primary schools. Eric Rogers was never happier than when he was talking to children. He did not ask them questions but told them stories, and would escape with them to the kitchen to try things out, not always endearing himself to their parents with the mess they made.
- A course should follow Eric Rogers' model of a curriculum driving towards clear and important end-points, embodying a connected web of knowledge. Understanding physics needs to be at the centre of the endeavour. Method and content of teaching need to respect the ways in which pupils learn.
- Examination questions must convey the message that understanding is essential. Cheap recall should be discouraged and equations be printed on the front of the examination paper. Marking should be humane, responsive to how the pupil answers, and respecting the teacher's judgment. Bonus marks should be available for candidates who do more than the examiner expects. Examiners need to be closely involved with teachers, ensuring a close link between what is taught and what is examined.
- Pupils need to play with equipment and ideas at a first meeting, knowing that there will be chances to understand better later on because the course will spiral round to them another time. First time around, enjoyment matters more than knowledge, because knowledge will not be gained without it.
- Changes in education work well when the teachers who take part are committed volunteers who put a lot of effort into making them work. Changes need to be tried out and revised before finally deciding on them. If changes are seen as being in pursuit of better education, morale will be high. The pupil's understanding of physics must be at the centre of the endeavour.
- Teachers need to be fully resourced and supported when they begin to implement changes. For me, help was no more than a phone call away, and support was provided at regular meetings.
- The authors need to be closely involved with the teachers and the examiners ensuring a close link between what is taught and what is examined.

In my list, the first two ideas seem now to be here to stay. Eric Rogers would have been pleased with the progress made in science in primary schools. But the rest, I have to say, seem to have been systematically ignored or denied in the recent period of change, with the introduction of a National Curriculum and changes in the way teachers are trained. That sad fact leaves me angry and frustrated. By stark contrast, I have tried to describe the humanity, thoughtfulness, commitment and involvement, and the respect and support for teachers, which Eric Rogers and others brought to their work.

You will see that I had the advantage of learning from the best of practitioners, in contact with the best recent thinking. That first briefing conference started me on the road of learning and teaching physics in the company of one of the most eloquent of teachers. We ought to ensure that future physics teachers have as good an introduction.

PERSONAL POSTSCRIPT

Eric spent much of the last years of his life in Cambridge, a place which was very dear to him, though after his wife died there in 1971, for some years he could not bear to return. When he did, I spent many an hour talking to him about teaching and physics. He would come up to the Cavendish Laboratory to play with apparatus and talk to students and children. Our regular dinners usually included physics conversations in which I knew I was being taught by the gentle arts of 'shredding' or story telling. He loved to teach and he was subtle about it.

Sadly I was almost there at the end. I arrived home late on the Sunday night of 1 July 1990 and Eric's voice on the answer-phone said, 'I need some dinner and some physics'. After work the next day I went round with cakes for tea. I arrived as Eric left on his last journey. He was alone when he died; a great sadness to those who had known him during his last years. All that remained to be done was to inform the world of physics education of the loss of one of its most remarkable characters.

Today's young teachers know little of how Eric Rogers' work laid the foundations of some of the best physics teaching in the world in the middle years of the twentieth century. His spirit will live on into the next century. *Physics for the Inquiring Mind*, translated into many languages, and his many dedicated 'children of physics', are his memorial.

PART 3

REMEMBERING ERIC ROGERS

Young people are thrilled with the idea of scientific experiments and knowledge. Many a small boy is eager to learn physics and chemistry. When we show him a plain test-tube, his tongue hangs out with enthusiasm. He longs to play with the first magnet he sees. Yet a few years of science courses—including, say, some qualitative analysis or a study of magnetic field formulae—will deaden the enthusiasm in almost all students. A few emerge still determined to be scientists—but even they usually have a strange picture of science as a sort of stamp-collection of facts, or else a game of getting the right answer. For the majority, well-meant teaching has built a wall around science, a stupid antagonistic wall of ignorance and prejudice.

Eric Rogers *Teaching Physics for the Inquiring Mind* p 2

CHAPTER 20

ERIC ROGERS 1902–1990

Keith Fuller

Bedales school, and its remarkable founder John Haden Badley, affectionately known to staff and pupils as 'The Chief', had a profound influence on the young Rogers, who was educated there from 1916 to 1921. Founded in 1893, Bedales began as, and remained, a revolutionary, progressive, co-educational boarding school set in a beautiful rural area near Petersfield, Hampshire.

Eric Rogers was born on the 15th of August 1902 at Bickley, Kent, the family home being at Crockham Hill near Chartwell. His father, Charles Knight Rogers, a publisher for the International News Co Ltd, sent him to Bedales in the footsteps of his elder brother and sister. He immediately felt at home, enjoying the robust outdoor life so beloved of Badley, and excelling at skiing and skating. He was active in Badley's famous theatrical ventures, and he became secretary to the lively Scientific Society which, encouraged by the renowned telecommunication engineers the Eckersley brothers, and aided by the Glasgow University Press, published an annual journal. Eric Rogers' interest in physics was nurtured by Archie Heath, a Bedales teacher of small stature, but possessed of a wonderful, eccentric intellect, who later became Professor of Philosophy at University College, Swansea. A contemporary, Francis Partridge, described Heath as '...a clever dwarf with a basso profundo voice and a furious temper when roused'. In his final year at Bedales, Eric Rogers was elected Head Boy, and Editor of the school magazine.

In October 1921, Eric Rogers went to Trinity College, Cambridge, gaining first class honours both in the Mathematics Tripos Part 1 in 1922, and the Natural Science Tripos Part 2 in 1924. Elected senior scholar in 1923, Eric Rogers also continued his theatrical interests at Cambridge, and maintained regular contact with Badley, attending Old Bedalian camps and accompanying 'The Chief' on walking holidays in Cumbria. A short period of teaching and research at the Cavendish Laboratory in 1924–25 ended with his first teaching appointment as physics master and as assistant house-master (Brown's House) at Clifton College, Bristol, where he remained from 1925 to 1928.

In 1928, Badley tempted him back to what Eric Rogers described as 'the jostling happiness of Bedales', offering the posts of physics teacher and boys' house-master. On some winter days, lessons were suspended in favour of skating and tobogganing, and Eric Rogers assumed responsibility for instructing a young history teacher, Janet Drummond, in the skills of skating. There followed what a friend described as an 'intense, passionate courting time'. Eric Rogers and Janet Drummond became engaged in April 1930, eloped from the school before the end of term, and married on the 14th of June 1930. The Bedales archive preserves a letter to Badley from Eric Rogers about these events:

A remark of yours, some years ago, made me resolve that if I married I would uproot myself from school and go into exile for a time. And the basis of the resolution I agree was right, though in carrying it out we were far too hurried (and owe many, many, apologies).

Bedales Archive

They took their 'exile' in the USA, where Eric Rogers became an instructor in physics at Harvard University for two years, conducting an active correspondence with Badley:

To my joy I was offered both some of the lecturing and an almost free hand in reorganizing one of the Harvard first courses in Physics (a general one for those who have done physics at school but will do no more after this). There is therefore plenty to do in eliminating mere logical substructure, and building up instead some idea of scientific ideas and aims and methods. I have been itching to be allowed to try this but little hoped to get it. It means lots of work, but is worth it.

Bedales Archive

Here are the seeds of Eric Rogers' long and fruitful interest in physics curriculum design, his famous 'block and gap scheme' for physics teaching, and his great book *Physics for the Inquiring Mind*. He built up his courses from carefully chosen end points and worked backwards from there to select the physics content.

His correspondence with Badley during this period contains vivid descriptions of American life and scenery, details of the birth of his son David in 1932, accounts of the many schools he visited and, of course, his unfavourable impressions of physics education in the USA. This was an important time for reflection and learning:

I have always wanted school to make people look forward to their occupation as a real pleasure, and grow up to enjoy it for itself. I cannot believe that any teaching should ever be done in any other frame of mind, anyway.

Bedales Archive

In March 1932, Eric Rogers wrote to Badley hoping to return to Bedales:

When I turn to Bedales it is no second or third choice, but my first choice far and away, anyway I want greatly to come back and do all I can for Bedales—with the same impatience to be working for you that made me leave research—I never regretted that defection.

However, his parents felt that he was not suited to the post, and unknown to him his father had written to Badley to suggest that Eric Rogers should not become a Bedales house-master again. Badley concurred, and his reply to Eric Rogers contained a 'stern refusal' about any future at Bedales, whilst encouraging him by pointing out his bright prospects.

Within a month of returning to England in June 1932, Eric Rogers had been offered no less than three posts, at Gresham's, Winchester and Charterhouse. He chose Charterhouse where he remained as physics master from 1932 to 1937. The Carthusian for 1935 describes one of his theatrical appearances:

Mr Rogers concealed himself very well under the vacant guise of a bashful and bewildered foreigner in a country of madmen.

In 1937, Eric Rogers used his USA contacts to secure an assistant headship at The Putney School, Vermont, remaining there for three years, followed by appointments at Mount Holyoke College (1940–41) and St Paul's, Concorde (1941–42). In 1942, he became associate professor (later professor) of physics at Princeton University, New Jersey, remaining there until his retirement in 1971. In the late 1950s he became a member of the Physical Science Study Committee, contributing to writing texts and making films.

In 1962, he was asked by Donald McGill, organizer of the Nuffield O-level physics project, to be a consultant to the project. His impact on the team was considerable, and upon McGill's untimely death in 1963 he was the natural successor, with expert help from John Lewis and Ted Wenham. After retiring from Princeton, he worked closely with Ted Wenham on the revision of Nuffield O-level physics. The death of Eric Rogers' wife in 1971, followed closely by the death of his sister in 1972, severely disrupted his retirement plans and, of course, work on the revision, causing some acrimony with the publishers.

Eric Rogers died alone in Cambridge on 1st. July 1990, aged 87. He was a brilliant, eccentric physicist who created a highly individual vision for physics education. He was frequently awkward and irascible, whilst at the same time being remembered as kindly, helpful and considerate. His long letters to Badley reveal strongly held views combined with great sensitivity. Badley's influence was lasting: chapter 24 of *Physics for the Inquiring Mind* is introduced by a quotation from him:

A time to look back on the way we have come, and forward to the summit whither our way lies.

> *We aimed for knowledge and a sense of understanding rather than a wealth of information... time for careful teaching and for students to learn by reading and thinking things out for themselves... We considered the loss of omitted topics unimportant. If such a course succeeds, its students will be well prepared, in both background and attitudes, to read more science on their own, to fill any gaps they may wish.*
>
> **Eric Rogers** *Teaching Physics for the Inquiring Mind: Introduction*

CHAPTER 21

REMEMBERING ERIC ROGERS

Anthony French

I spoke in my ICPE address (see p 229) of the electrifying effect on me of first hearing Eric Rogers speak. A lecture to a sizeable audience, enriched and enlivened with demonstrations, was undoubtedly Eric Rogers' natural way of teaching physics. As anyone who ever heard him knows, he was a superb showman, with a good deal of the ham actor in him. He took mischievous pleasure in challenging, perplexing and astonishing his public. In presenting him for the award of the Oersted Medal of the American Association of Physics Teachers in 1969 (reprinted on p 227), the late Walter Michels praised him for his showmanship, his sly and unexpected turns of phrase, and for his deep insights into students' minds.

The great memorial to Eric Rogers' life and work is, of course, his book *Physics for the Inquiring Mind*. Into it he put the fruits of his decades of creativity and thought about presenting physics as a living and exciting subject. Although his audience consisted of students who were not going further in physics, the book contains more intellectual stimulus for the future professional than do most of the standard textbooks. His intensely personal flavour comes through in everything he published; he wrote as he spoke.

Eric Rogers would surely have agreed with Samuel Johnson that, 'I know nothing that can be best taught by lectures, except where experiments are to be shown.' In the last talk I heard him give, at a GIREP conference held in the Netherlands in 1984, he began by claiming that Poincaré, the great mathematical physicist, is said to have remarked, 'The British are very strange; they use experiments.' He proceeded to present a dazzling sequence of demonstrations, including a typically provocative Rogers twist in which the motion of a pendulum (with an iron bob) was distorted by a hidden electromagnet operated by someone behind the scenes. His aim was always to keep the audience on their mental toes, sometimes by mystifying them and then triumphantly resolving the mystery. It is a tribute to his mastery of the process that he could work this magic not only with students, but also with his fellow teachers. Every physics teacher should read his article, 'A Talk about Demonstrations' in the book *Physics Demonstration Experiments*, edited by Harry F Meiners.

In 1956 Jerrold Zacharias got Eric Rogers involved in the PSSC physics films. This brought his powers as a master demonstrator to tens of thousands of students who would never otherwise have seen him in action. He made only two films, as far as I know: *Coulomb's Law* and *Coulomb Force Constant*. Both are imbued with his ingenuity and his delight in dramatizing the display of physical phenomena. The film on Coulomb's Law in particular was the impudent Rogers at his most characteristic.

It was not my good fortune to know Eric Rogers well, but it seemed to me that his extraordinary effectiveness as a teacher came from a combination of several factors. First, an almost childlike delight in the wonders of the natural world, and a freshness in approaching it that never became stale. Next, the acute and cultured intelligence with which he analysed our interaction with that world. His love and knowledge of art and literature can be seen in all his work. Then, the inventiveness and imagination with which he devised ways of communicating his insights to others, helped by great eloquence, impish humour, and a rare gift for the memorable phrase. And, last but by no means least, the ebullient personality, and the self-assurance, with which he commanded our attention. As Jerrold Zacharias, a rather similar personality, used to say, *It is all right to be a prima donna if you can really sing*! I feel lucky to have experienced Eric Rogers' act at first hand.

Our courses should mediate between the scientist and the layman, between a classical culture and a scientific civilization. They cannot do this just by pouring in scientific information or even formal training. What is needed is a genuine understanding of science and the way scientific work is done. To make this understanding a lasting part of peoples' culture is a huge task. In a year's course we can only give a glimpse of it.

Eric Rogers *Teaching Physics for the Inquiring Mind* p 9

CHAPTER 22

ERIC ROGERS IN DENMARK

Poul Thomsen

I met Eric Rogers a long time before he met me. More than 30 years ago, when I was teaching at a Danish teacher training college, I used parts of the newly developed PSSC physics course, including some of the PSSC films. One of them, which I immediately fell for, had the title *Coulomb's Law*. In it I saw a small, friendly-looking man doing experiments on Coulomb's law with an ingenious balance. Not only did I like his experiments, I was fascinated by the way he talked and gesticulated with such enthusiasm that nobody could doubt that here was someone who both loved physics and loved to share this love with others. This man was Eric Rogers.

Several years later, at a meeting at Malvern College, England, in 1968, we met face to face, and began many years of friendship. In that year we arranged a seminar in Copenhagen on the teaching of physics in schools. Eric gave a lecture on the 'Educational theory of teaching conservation of energy', and since I spoke about experiments for teaching energy, we had several fruitful discussions during the conference.

Later, I visited Eric Rogers in Princeton. I remember walking outdoors with him on a cold and icy January day. He had no overcoat, and I asked him if he should not find one. He said: *No, I don't need it, I am water- and fire-proof*. Back in his large lecture room, Eric told me how he loved to combine theory with experiments which would shock the students and prevent them from falling asleep. Once he got the students to do calculations about a ladder against a wall. Theory predicted that the ladder would slip if someone climbed up it beyond a certain rung. A tall ladder was set up and Eric climbed up it. As predicted, when he reached the critical rung, the ladder slipped, frightening not a few of the students. What they did not know was that he had hidden a safety stop behind the demonstration bench, so that the ladder only slipped a small distance.

Eric Rogers was an excellent speaker at international conferences on physics education, and was always willing to give one of his inspiring lectures whenever asked to do so, so he was very much in demand. He always travelled light, with the absolute minimum of luggage. He told me that he carried a card stating that he held a professorship at Princeton, and that he always showed this card when he arrived at a hotel almost without luggage. *If I do not show this card, they think I am a tramp*, he said, *but as soon as they see the card they treat me very politely and do not ask me to pay for my room in advance.*

Shortly after my return to Denmark from the USA, I met Eric again at a conference in Eger in Hungary in Autumn 1970. There I discovered that, like me, he was a keen swimmer, so we often went to the thermal baths in Eger, discussing physics both on the way, and even in the pool!

At that time, Eric became director of the English Nuffield physics project and spent much time in England, often coming to Copenhagen to spend some days at our Physics Institute, giving public lectures on physics education with demonstrations of new experiments, which he performed in his usual inspiring way. He always started rather formally with his jacket nicely buttoned. He insisted that teaching physics in a dull and boring way was bound to lead future parents to have negative attitudes. I remember one of his characteristic overhead projector sketches, in which a mother says to a child who is just starting physics: *Do not expect too much from it. It's dull. I hate physics.*

He wanted physics to be taught in a much more inspiring way, by letting the students learn by doing experiments, and by observing other experiments done by the teacher. So, in these lectures, he was soon eagerly engaged in doing experiments. He would take off his jacket and throw it in a corner on the floor. Sometimes he would jump up on the table. Once I saw him walking around on the tables amongst the audience, with a flask of solid carbon dioxide. All these activities were a disaster for his clothing. His shirt would come out of his trousers, which in turn would slide a little downward, so that he often had to pull them up and tighten his belt.

Eric cared little for appearances. Once, when we were dining in a restaurant, he promised to get some papers for me, and without hesitation tied a large knot in his necktie. *It's a very good way to make me remember it*, he explained. *When I cannot get the tie over my head tonight I will know I have to remember something and I will write it down.* When we left the restaurant he did not care a hoot about the strange looks he received.

Whenever Eric gave a lecture at our physics institute we invited members of the Danish physics teachers association, and they came in large numbers. In this way he had a considerable influence on their teaching of physics. At that time we were preparing a reform for the Danish Folk School, the public school for students of age 7 to 16. We wanted the teaching of physics and chemistry to be based, much more than before, on students' own work in the laboratory. Three outstanding physics teachers and I worked hard on preparing new teaching units, which we tested and revised in a large number of school classes. Our work became known as the Danish 'Ask Nature' project. Eric was deeply interested in the 'Ask Nature' project, just as I was in his work with the Nuffield project, and we had a lot of fruitful discussions. I am very grateful for his help over those years.

When Eric Rogers visited my family in Copenhagen, our youngest daughter, Lene, took him to her heart as an extra grandfather. He brought her gifts and they had a lot of fun together. Once when they were sitting together in a nearby room, my wife and I heard Eric reading loudly in Danish with a heavy English accent, with a lot of laughing. Listening carefully we realized that he was reading from a Danish Donald Duck magazine. They were playing a game, for which Eric was mainly responsible, in which he had to read aloud in Danish, look at the cartoon and guess what he was reading.

This little episode illustrates how good Eric Rogers was with children as well with adults. Another illustration is the way Eric noticed that we had a brilliant laboratory assistant who could do practically anything with his hands. It is unbelievable what Eric got him to do to produce pieces of equipment for his demonstrations.

We had long discussions about tests and examinations. It was very difficult for me to convince him that oral examinations, which we often use in Denmark, are suitable for testing knowledge and understanding of physics. During one of his visits, I was to act as external examiner at a teacher training college in Ålborg in Jutland, and he accepted my invitation to come and see how these oral examinations worked in practice.

Let me describe how we conduct these examinations. The student draws by lot an examination question about how to plan and teach a physics topic at a specified level, and has three hours to prepare an answer followed by half an hour to present it. The student is expected to explain the choices made, and to demonstrate the ability to do the experiments. The examiners may ask questions during and after the presentation. Then another question is drawn, this time a theoretical one, aimed at testing basic knowledge of physics.

As he watched, Eric Rogers became keenly interested, and when the teacher and I were discussing what marks we might give, he asked if he might give his own evaluation before we told him our decision. It turned out that he was very well able to distinguish between good and less good students, usually getting quite close to our marks despite not knowing the language. I believe that he felt more positively about oral examinations after this event, but he still said that it would never be possible to introduce examinations of this type in England.

In 1988 we began a new physics curriculum for the Danish Folk School. I think Eric Rogers would have been pleased to know that we reduced the teaching of classical mechanics to make room for astronomy. He had always tried to convince me that we placed too much importance on mechanics.

Now I am again engaged in writing schoolbooks, trying to live up to the new intentions in this curriculum. I regret that Eric Rogers is no longer among us. I would have liked very much to discuss our new approach with him. Nevertheless he is still an influence on the way we are writing the new books. I know for certain that the many hours I have spent with him have had a profound effect on my thinking.

Scientists feel driven to know, know what happens, know how things happen; and, for ages, they have speculated why things happen. That drive to know was essential to the survival of man—a generation of children that did not want to find out, did not want to understand, would barely survive. That drive may have begun with necessity and fear; it may have been fostered by anxiety to replace capricious demands by a trustworthy rule. Yet there was also an element of wonder: an intellectual delight in nature, a delight in one's own sense of understanding, a delight in creating science. These delights may go back to primitive man's tales to his children, tales about the world and its nature, tales of gods. We can read wonder and delight in stone-age man's drawings; he watched animals with intense appreciation and delighted in his art. And we meet wonder and delight in scientists of every age who make their science an art of understanding nature.

Eric Rogers *Physics for the Inquiring Mind* p 211

CHAPTER 23

ERIC ROGERS: A PERSONAL MEMORY

Goronwy Jones

I well recall my first meeting with Eric Rogers, now with feelings of fond amusement, though at the time the impact was rather different. I went to Loughborough, to a briefing conference on all three Nuffield science projects. I had of course read his *Physics for the Inquiring Mind* and enjoyed it. I had written to him some months before asking if he could invite a local school to take part in the physics trials. I felt that with so much mystery surrounding the project, to have a local school taking part would lessen the mystery and an attitude which bordered on hostile curiosity. He wrote back in generous terms thanking me for my 'magnificent' letter, but said the money to equip trial schools had been exhausted, and it brought me an invitation to Loughborough. When an H.M.I. offered to let me read the project material over 36 hours, I worked like a demon to make my own very brief précis. I found the guides good reading and, having only recently left the school classroom, I could not wait to go to the conference.

Arriving at Loughborough, I went to register and was joined at the desk by the great man. I murmured that I had managed to steal a look at the books and had made some notes in preparation for the conference. The next few minutes I shall never forget. Eric Rogers exploded with indignation and fury. It went on for two or three minutes but seemed much, much longer. With threats to sue if I did not immediately burn my notes ringing in my ears, the only way to escape this manic storm was to flee to my room.

The rest of the conference, which I had approached with delight, became at least in parts a nightmare. I enjoyed the practical sessions but I was never relaxed. The 'shredding sessions'—a new experience in itself for most people—became an ordeal as bad or worse than a viva voce examination with a hostile, threatening tutor. I drew little comfort from learning that others had received similar rough treatment. A media producer had sought an audience and reported at the appointed hour. Eric was holding a shredding session and would not be disturbed. As time went far past the appointed hour,

217

the producer sent word to say he would have to leave in 10 minutes to catch the last train. Back came the furious instruction that he was to wait or else. He waited.

The fraught times in which he led the project, difficult and unnerving for many people, contrasted totally with my later meetings. I recall his antics on the lecture stage at Edinburgh, demonstrating a simple way to show the parabolic flight of a projectile. This was not manic; this was a leprechaun entertaining us and making fun with physics.

Another time, with his audience on the edge of their seats as he crackled along with his Maxwell Demons, a wag asked: 'How do they know which way to push?'. 'They're always against the government—they're Irish you see', came the instant reply. My Irish friends chuckled as much as the rest of us.

Later at a GIREP conference at Oxford on 'The role of the laboratory in physics education' I saw him working hard in a darkened lecture theatre— he was now an old man—setting up an arrangement to show dark field microscopy. He used ordinary school laboratory apparatus which encouraged many present to try for themselves.

On all three occasions he was as busy as ever, but there was a more relaxed humour still richly laced with wit. We could approach and ask for details and engage him in general conversation, without the trepidation of my Loughborough experience.

In time, a revision of the Nuffield physics project became necessary. That this proved as successful as it did was largely due to Ted Wenham (figure 14.1). Ted knew that Eric could not expect the indulgence he had previously been granted. When Eric had made some unfortunate points in a lecture at Loughborough, Ted stood up and rebutted him in terms that everyone respected. I am sure that Eric was thankful that he had a colleague who was such a friend. When the revision was planned, Ted, who knew so much about its clientele, was crucial to its tempo and temper.

Years later I met Eric Rogers at several conferences abroad. This was a mellow man with whom one could chat, discuss and disagree. When his distinction was recognized by ICPE with their medal for physics education at the Trieste conference, his pride was obvious but so was his modesty. He seemed to want people to think he was ordinary, perhaps so that he could surprise them. However, he was growing older and the adrenaline flowed less freely. I was pleased for him when he spoke of work he intended to do with the interactive kind of museums that are now so fashionable. He spoke of needing an arrangement for youngsters to produce a small rainbow and possibly a larger demonstration one. I had some years earlier used an arrangement designed by Colin Siddons using small polystyrene spheres and a small 2.5 volt lamp. I was flattered by his interest and sent him samples.

His vision of physics education was not, in my opinion, as fully realized or appreciated in the 1970s as it should have been. 'Stage managed heurism' was what Bill Ritchie (figure 14.7) called it and, had more had the courage of this conviction, there would be far less of the turmoil we now have in UK curriculum development. Certainly having taught the Nuffield physics project and marked the scripts of 16-year-old candidates whose understanding of difficult principles often exceeded that of graduates, I still applaud his ambitious expectations for young people learning physics. That the Nuffield Physics project of the 1960s has had an influence cannot be in doubt. One has only to read recent textbooks and syllabuses to appreciate it, though the influence on classroom tactics and strategy may be less evident. The apparatus designed by the Project was a vast improvement. Previously much equipment was 'hand-me-downs' from advanced level or even undergraduate courses. Now 11- to 16-year-old pupils could make simple direct observations uncluttered by the refinements needed later. As Bill Ritchie said 'It did not guarantee success, but it made it more likely'.

That Eric Rogers, whose contribution will be valued for many years to come by those who love physics education, should go to his last resting place so little noticed I shall always recall with sadness and even guilt. In his last years when he retired to Cambridge, Brenda Jennison helped him enormously. He was no longer able to walk without discomfort and Brenda not only chauffeured him to places he wanted to visit, but kept him involved in his beloved physics. She invited him to meetings of her Physics Centre and the Physics at Work sessions at the Cavendish Laboratory. Indeed it was at one of the Physics at Work exhibitions, that Brenda organized so brilliantly, that I last saw him. We chatted as he sat huddled away, while pupils and students from the local schools and the University milled about, little realizing that the enjoyment they gained from their education through physics owed so much to this little grey-haired man.

If this volume achieves nothing else it may serve to remind those who knew him and others who did not, how much is owed to the imagination and vitality of Eric Rogers.

Having gathered my staff of class teachers, I say to them:

'Your students are non-scientists. They are well-educated, intelligent young people, but this is their final science course—that is, unless we leave them so delighted that they want to go further. This is their only chance to meet and talk with a good physicist face to face. So please respect their interest and their questions. If they raise a question, chase after it, answer it, and encourage discussion. If a chance remark leads to something interesting in your own field of work, let yourself go, dive in with enthusiasm, talk, explain and discuss. Never mind if this plays havoc with the syllabus. In this course, topics can wait, or be shelved forever. The important thing is to have the students understand science and know and like scientists.'

Eric Rogers *Teaching Physics for the Inquiring Mind* p 23

CHAPTER 24

ERIC ROGERS IN BRAZIL

Susana de Souza Barros, Marcos da Fonseca Elia, Rachel Gevertz and Ana Tereza Filipecki Martins

24.1.   HELPING STUDENTS UNDERSTAND PHYSICS

Susana de Souza Barros

This chapter contains contributions from several generations of physics teachers who have tried to bring alive in our country what the learning of physics can offer to the inquiring mind. We present the thoughts and recollections of a number of physics teachers who, personally or indirectly, were touched by Eric Rogers' wisdom.

Eric Rogers' influence in Brazil goes back a long time. In 1963 he gave a talk, *Should we teach physics for information or understanding?* to the First Inter-American Conference on Physics Teaching in Rio de Janeiro. He argued, as always, that without understanding there will be little left of facts and information. Many of our small groups of Brazilian idealists are still trying, thirty years later, to put that message into action, through work in the classroom and through training and re-training teachers.

While living in the USA during the 1960s, and before I had heard of or met Eric Rogers, I began to become aware of the difficulties students could have in learning physics. I met many kinds of students: science and technology students at a well known university, students who were training as elementary school teachers, and minority students given special training in their last two years of high school to help them get to university. This led me to search for the basic ingredients necessary to understand physics. I came to understand that people do not learn only by listening or reading, or even by doing. They learn when they are cognitively and emotionally ready.

My search also brought me into contact with many interesting books, since that was the era of the big curriculum development projects.

I then learned that behind more than one of these projects there was an English professor teaching at Princeton, who had written a very special text for non-science students: one Eric Rogers. I found *Physics for the Inquiring Mind* and fell in love with it at first sight. It gave me many hints of how I might start answering some of my problems. One insight was that to help students to learn physics we need to take them on a guided tour through early scientific notions; that astronomy was therefore not to be forgotten.

I also began to understand more fully from it the importance of practical observation of phenomena and how to understand a model or theory, and I resolved that my courses would give more phenomenology and less content. For example, starting on the kinetic theory of gases, I would now want to have students notice how long it takes for the smell of different volatile liquids (ammonia, alcohol, perfume) to reach their noses across different distances, after the bottle is opened, and to compare this with diffusion in liquids. After these and other observations, I would want to ask whether the results are best explained by thinking of a gas as many small particles moving quickly, or as a substance expanding through the space. The student needs to think in terms of emptiness and the facility of squeezing through space of whatever comes out from the bottle. Only after this kind of experience of phenomena and some imaginative thinking would I want to begin to work out equations.

Similarly, one way to start thermal physics is to ask the students to say whatever they can think of about heating and cooling, how it is done, what does it, and how it works. This list can be very simple and down to earth but it contains their own ideas, associations and knowledge: stove, refrigerator, gas, water, Sun, iron, light bulb, coal, ice, cover, fan, to boil, to cool, thermometer, etc. Discussing this list, ideas related to heat can be introduced: conduction, convection, expansion, transforming mechanical energy into heat, condensation, evaporation, temperature scales, fixed points, pressure, etc. In this way, most of the concepts can be dealt with, and many important properties and phenomena can be discussed, without at first treating them in a formal and perhaps sterile way.

Before ever meeting Eric Rogers in person, I became acquainted with some of his Princeton students. For family reasons I used to visit Princeton. More than once I met Princeton University students on the bus from Kennedy Airport, and I always asked about their science courses, hoping to learn how well Eric Rogers succeeded. I wanted to know if the author of *Physics for the Inquiring Mind* was able to get as much as he thought from his students. It was remarkable to find that these girls and boys could understand the beauty of a science like physics, because somebody like Eric Rogers was there to teach them. It was a real pleasure to talk to them, and seldom did I get uninteresting or uninterested answers to my unfailing curiosity.

24.2.   LEARNING FROM PHYSICS FOR THE INQUIRING MIND

Ana Tereza Filipecki Martins

I am a secondary physics teacher in a Brazilian technical school. Although I never had the pleasure of meeting him personally, since I became acquainted with Eric Rogers' book *Physics for the Inquiring Mind* I have found in it answers to some of my difficulties, which were especially helpful to me in preparing and thinking about my teaching.

His approach to electricity made me change completely how I treat the subject. I decided to start with current electricity instead of the usual electrostatic approach. I also divided students into groups and asked them to describe their observations and to write down their ideas, before I introduced any formal concepts. Experimenting and thinking for themselves like this got my teenage students very enthusiastic.

Students initially discussed their ideas within the group and, afterwards, in the whole class. This thinking about simple qualitative and semi-quantitative experiments helped them when I introduced analogies such as a water circuit and an electrical 'pressure gauge'. And it improved their performance on tests of their comprehension, which also allowed me to understand their difficulties better.

What makes *Physics for the Inquiring Mind* so special? It is a combination of clever arguments, simple and easily performed experiments, very well constructed exercises and plenty of creativity. The way ideas are presented makes us re-think concepts we had thought we understood, and forces us to re-shape them again for ourselves. That aspect alone gives a very special place to Eric Rogers' contribution to the continuous self-training of a physics teacher.

24.3.   LEARNING FROM ERIC ROGERS

Rachel Gevertz

I first met Eric Rogers in Brazil in 1971, when he gave a course on tests and test construction for physics teaching. Having earlier myself introduced the PSSC course in Brasilia, and having previously at Harvard met some leaders of PSSC such as Jerrold Zacharias, Uri Haber-Schaim and Philip Morrison, we had mutual friends and common interests and had much to talk about. I told him about the work of the Science Teaching Centre of the North East but our conversations went far beyond evaluation and PSSC to reach deeper into Brazilian problems, particularly about the North East.

Eric Rogers was the first to awaken in me an understanding of the importance of testing all activities related to the teaching-learning process. I learnt from him that testing can be a very powerful instrument to improve teaching. He insisted that one should begin any teaching by thinking about a process of continuous evaluation, and by preparing tests for what you are going to teach. He helped me to see that this was also a way of discussing the purpose and objectives of physics teaching. Following his advice in work in schools in São Paulo helped bring efficiency and quality to the renewal of physics teaching.

On the last day of our meetings, just as he was leaving the room, he turned to me and asked, 'Your colleagues from the North East, do they really need physics?', and moved away without giving me the chance to answer. I was taken so much by surprise that I did not have an answer at the time. It is a difficult question with a still more difficult answer. Reflecting on his question, I came later to see what he meant. Had I been so taken over by my enthusiasm, working with teachers in remote regions of Brazil, that I was ignoring their real needs, needs that should permeate all educational action?

Two important messages from Professor Rogers, which we kept in mind throughout the years and of which educators should always be reminded:

Test to teach; do not teach to test.
Understand the real situations and needs of others, and plan accordingly.

24.4.   LEARNING FROM 'SHREDDERS'

Marcos da Fonseca Elia

When writing my doctoral thesis about examinations in physics, I read about Eric Rogers' 'Shredding Sessions' in the report of his 1970 Montevideo Seminar. Although such a session may look like a conventional pre-examination meeting to discuss questions, it is not. Its purpose is very special:

> *The process of constructing questions and criticizing them soon leads to an informal discussion of the philosophy of science teaching. It is this last that is the most important outcome of such a seminar. We all have some philosophy of teaching: ideals, ideas and methods, hopes; concepts never fully formed, never fully recognized, yet with profound influence on our teaching.*
>
> *Let me listen to your examiners making their examination questions and let me listen again when they are marking the answers, and I shall know the real value and promise of your work.*
>
> **Eric Rogers**: *Montevideo Seminar UNESCO* 1972

I had the opportunity in Trieste in 1980 of participating in one of these sessions led by Eric Rogers himself, and was very impressed by the discussions we had and by the potential consequences they could have in my teaching, if I could follow the same process with my colleagues. For me these sessions seemed like a kind of therapy, to sharpen the mind and sensitize feelings.

Back in Brazil, I tried to spread these ideas as much as I could. For instance, as part of a physics examinations course I gave for high school teachers during the 1985 Brazilian National Symposium on Physics Teaching, I invited the great and influential French–Brazilian physics educator Professor Pierre Lucie to be the moderator (Eric Rogers' term was 'grandfather') of a shredding session. I need hardly say that it was a highlight of the course. And in this way I brought together in spirit two of those who have most influenced my professional life.

Today, more than 20 years since Eric Rogers' words were written and 10 years since I came across them, his words and ideas maintain an invisible but real influence on my daily physics teaching and research work.

CHAPTER 25

THREE MEDAL CITATIONS

ERIC ROGERS
OERSTED MEDALLIST
1969

American Association of Physics Teachers
Citation by Walter Michels

'Eric M Rogers of Princeton University stands before us as one of the few indubitable violations of the Law of Conservation of Energy. For the past quarter century there have been few AAPT meetings that have not been enlivened by his contributions of new ideas for elegant demonstration experiments, or graced by his subtle insights into problems of clarifying physical concepts for both science and non-science majors. Volumes of the *American Journal of Physics* contain numerous pages of his authorship. *Physics for the Inquiring Mind*, based on his famous course at Princeton, stands as a landmark among texts that convey a lucid, carefully correct development of the subject matter of physics, together with mature and sophisticated discussion of the intellectual facets of science—its origins, methods, glories and limitations. His handiwork and turns of phrase are to be found in both the text and films of the PSSC course for secondary schools. His leadership of the Nuffield Foundation Physics Teaching Project is having a forceful impact in Great Britain, and is now percolating to the United States and elsewhere around the world. The stamp of his ideas and insights is visible in the records and output of more national and international conferences on physics education than it would be practicable to mention. His vivid demonstration lectures at the Franklin Institute, at the Royal Institution, on television, and as a visitor in schools and colleges, have delighted and instructed countless individuals both old and young.

One can enter his lecture room and find him non-conserving energy by climbing a rope to the ceiling, or walking on his hands, or pouring water into a suspended rubber bag until it bursts in a gorgeous flood, accompanied by an indelible impression on the by-then breathless audience. Combined with the showmanship and the sly and unexpected turns of phrase, are deep insights into the learning obstacles, conceptual and linguistic, that must be overcome in student minds; there are also the equally deep insights into how to write problems and examinations that focus student attention on the important aspects of learning: process and reasoning, relationships and values, and carry over to new situations, as opposed to a focus on routine or desiccated end results.

Many of his students and teaching assistants are now themselves engaged in teaching, and one can almost invariably detect the Rogers influence in their stance, their love of what they are doing, and in the values they convey to their students. Eric Rogers has always been a living example that neither science nor scientists need be 'square'—to use the current jargon. Throughout his text, his films, his articles, and his talks, he has always conveyed his sense of humour and his zest and love of physics. He has been the very opposite of 'square'. To paraphrase a now-famous rhetorical question from a certain film on Coulomb's law: 'Well, what would you expect of an inverse square? Law?'

For the example he has set us as a teacher, for the many materials he has put into our hands to help us improve our own work, the American Association of Physics Teachers awards Eric M Rogers this 1969 Oersted Medal.'

American Journal of Physics 1969
37 (10)
ⒸAmerican Association of Physics Teachers

ERIC ROGERS
PHYSICS EDUCATION MEDALLIST
1980

International Commission on Physics Education
Presentation Address by Anthony French

'Ladies and Gentlemen,

Today I have one of the most pleasant duties that has ever fallen to me as Chairman of the International Commission on Physics Education.

About one year ago the Commission decided that it would be appropriate to establish a medal to honour individuals who had made outstanding contributions to physics education on an international scale. In instituting such a medal we would be parallelling various awards for internationally recognized contributions to research in particular fields. It was agreed that the first award of this medal would be made on the occasion of the joint conferences at Prague and Trieste this year.

When it came to considering who should be the first recipient of the medal, the Commission arrived very quickly at the name of Professor Eric Rogers.

Perhaps I may be permitted a word of personal reminiscence here. Although the name Eric Rogers had long been familiar to me, I did not hear him in person until about 1960, when he spoke at a meeting of the American Association of Physics Teachers in New York. He must, I think, have recently been reading Norwood Russell Hanson's book, *Patterns of Discovery*, and he used an illustration from it as a jumping-off point for his talk. The effect on me was electrifying. I expect many of you have had similar experiences. I had never heard physics teaching talked about in such a provocative and exciting way.

Professor Rogers has spent a whole professional life in teaching physics and in advancing physics education. He had his initial education in England and obtained his baccalaureate degree at Cambridge University in Mathematics and Physics in 1924, followed a few years later by the MA. For a short time he worked under Lord Rutherford at the Cavendish Laboratory, Cambridge, but then decided that what he really wanted to do was to teach. He taught physics at a number of schools and colleges in England and the USA, and finally found his permanent home at Princeton University in 1942; he has been attached to that university ever since.

For many years his chief interest was in the production of physics demonstrations, which he wove into his teaching. All of us must know his wonderful book, *Physics for the Inquiring Mind*. It is, in the best sense, an idiosyncratic book. One can hear and see Eric Rogers on every page as one reads it.

More recently, he has turned his attention to testing, which does so much to dictate what happens in physics courses. His 'shredding' approach to testing is no doubt well known to many of you.

In addition, he played a major role in the Nuffield O-level physics project, from its beginning in 1963 until he retired from the project a year ago. He has lectured and conducted workshops, etc., all over the world. I expect many of you have heard him and taken part in these activities.

All of this fails to convey the force of the personal impact that Eric Rogers brings to everything he touches. But we shall experience that for ourselves in a few minutes.

A few words about the medal itself. The Commission considered and rejected the idea of naming the medal after a particular physicist, and making it in his or her image. The international character of physics teaching seemed to make this inappropriate. Instead, the distinguished Hungarian artist, Miklós Borsos, was given a free hand to make a more symbolic design. Mr Borsos exhibited his sculptures a few years ago at the Art Biennale in Venice.

The artist created a design showing the interaction of human beings with the forces of nature. He shows the four elements of the Greeks; Earth, water, air and fire—this last symbolized by a powerful ray of sunlight, like a sword. People are at first afraid of these forces, but then begin to make use of them—we see one person lighting a fire.

Professor Rogers—Eric—you have yourself lighted intellectual fires in the minds of thousands and thousands of students and thousands of physics teachers throughout the world. The Commission feels honoured to be able to give you this token of recognition and esteem.'

Proceedings of the International Conference on Education for Physics Teaching 1980 Trieste pp 80–81
© ICPE 1981

ERIC ROGERS
BRAGG MEDALLIST
1985
The Institute of Physics
Citation

Professor Eric M. Rogers was awarded the Bragg Medal and Prize for his many contributions to physics education, both in the USA and the UK, through his lectures, writing and work on both the Physical Science Study Committee (PSSC) and Nuffield teaching projects. An outstanding example of his work is the success of *Physics for the Inquiring Mind* (1960 Princeton University Press). This 780-page volume, stemming from his long experience as a teacher and resting on his conviction that 'educated people need some lasting understanding of physics', is still popular, sales nowadays being greater in the UK (and Hong Kong!) than the USA, where it was first successful. The book, which comprises the general reading, problems and laboratory instruction of his physics course at Princeton for non-scientists, was brought out with support from several foundations as an educational document showing a working programme of 'teaching physics for understanding', in contrast to the traditional mechanical memorizing. It was providing briefing material for UK physicists starting work on the Nuffield Science teaching project even before Eric Rogers joined the latter.

His connection with the Nuffield Foundation began in 1963, when on sabbatical leave in England. After visiting early meetings, by invitation, he joined as a consultant; then, later, he was asked to become Organizer in charge of the Physics Teaching project. He was loaned by Princeton to see the project through to publication; the five volumes of *Physics Teachers' Guides* which he compiled and others he edited have been in use in many schools throughout the UK ever since.

His other writings include many published papers on the aims and methods of science teaching, including his 'Block and Gap course in physics' (*Am. J. Phys.* 1949). Professor Rogers has also lectured widely. He gave the 1979–80 Royal Institution Christmas lectures on 'Atoms'—which were televised in colour by the BBC. He has also made films while with the PSSC, centred at the Massachusetts Institute of Technology. Honours he has received include an honorary DSc from Wooster College in 1968, the Oersted Medal of the American Association of Physics Teachers in 1969, and the first medal of IUPAP's International Commission on Physics Education in 1980. He is a Fellow of both The Institute of Physics and the American Physical Society, and a member of the European Physical Society. His other interests include the sociology of examinations: he advocates special seminars, which he calls 'shredders', to manufacture questions for examinations or tests as a device for training teachers in new teaching.

Physics Bulletin 1985 **36** (4) p 153.

Yet there were practical needs, in medicine and navigation, to keep good science alive. Then came the cry with growing fervour, 'Watch what does happen; stop arguing about what ought to happen'. Prejudice was being pushed aside by careful thinking in terms of experimental observations.

Eric Rogers *Physics for the Inquiring Mind* p 287

Again he was overjoyed at wresting a divine secret from Nature by brilliant guessing and patient trial.

Eric Rogers *Physics for the Inquiring Mind* p 269

CHAPTER 26

THE EQUIVALENCE PRINCIPLE DEMONSTRATED

Eric Rogers and I Bernard Cohen

26.1. THE PROBLEM

Eric Rogers

While I was living in Princeton, my wife and I would from time to time take a small puzzle involving physics to our neighbour Professor Einstein—often as a birthday present.

The last of these, presented on his seventy-sixth birthday, was, I believe, original. It was derived from an old-fashioned toy for small children: a ball on a string is tied to a cup in which the child has to catch the ball. But our modification was for Einstein a problem which he enjoyed, and solved at once.

A metal ball attached to a smooth thread is enclosed in a transparent globe. There is a central, transparent, cup in which the ball could rest; but initially the ball hangs by the thread outside the cup (as shown in the diagram). The thread runs from the ball up to the rim of the cup and down through a central pipe. Below the globe the thread is tied to a long, weak, spiral spring protected by a transparent tube which ends in a long pole—a broom-handle.

Starting with the ball hanging down, get it into the cup by a 'sure-fire' method.

The boundary conditions and information:

(1) the globe and the transparent tube should not be opened,

(2) the ball is made of solid brass,

(3) the spring is already stretched, in a state of tension, even when the ball is in the cup; but it is not strong enough to pull the heavy ball up into the cup,

(4) the broomstick is long,

(5) there is a method which will succeed every time—in contrast with occasional success by random shaking.

233

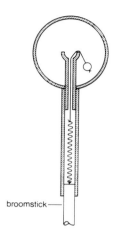

Figure 26.1.   *Einstein's birthday present*

And

(6) for readers here, a relevant reminder: it was made as a present for Einstein—who solved it in a real experiment, with delight.

26.2.   THE SOLUTION

I Bernard Cohen

...At last I was taking my leave. Suddenly [Einstein] turned and called 'Wait. Wait. I must show you my birthday present.'

Back in the study I saw Einstein take from the corner of the room what looked like a curtain rod five feet tall, at the top of which was a plastic sphere about four inches in diameter. Coming up from the rod into the sphere was a small plastic tube about two inches long, terminating in the centre of the sphere. Out of this tube there came a string with a little ball at the end. 'You see,' said Einstein, 'this is designed as a model to illustrate the equivalence principle. The little ball is attached to a string, which goes into the little tube in the centre and is attached to a spring. The spring pulls on the ball, but it cannot pull the ball up into the little tube because the spring is not strong enough to overcome the gravitational force which pulls down on the ball.'

A big grin spread across his face and his eyes twinkled with delight as he said, 'And now the equivalence principle.' Grasping the gadget in the middle of the long brass curtain rod, he thrust it upwards until the sphere touched the ceiling. Now I will let it drop, he said, and according to the equivalence principle there will be no gravitational force. So the spring will now be strong enough to bring the little ball into the plastic tube. With that he suddenly let the gadget fall freely and vertically, guiding it with his hand, until the bottom reached the floor. The plastic sphere at the top was now at eye level. Sure enough, the ball rested in the tube.

With the demonstration of the birthday present our meeting was at an end.....

(From 'An interview with Einstein'
Scientific American 1955 **193** July pp 69-73.)
From: Einstein *A Centenary Volume* ed A P French
1979 pp 131–132, Heinemann for ICPE.

Judging, in business or in government or in science, is not an easy job to be done by a careless, stupid guesser. It needs cunning and skill as well as a wide range of experience and educated knowledge, and it needs a ruthless spirit. Watch for opportunities for judging in your own work, now in your studies and later in your profession. If you rise high in your profession you will certainly do much judging—skill therein is a prime quality of good administrators. Rightly used, judging, with its rough answers, is good science.

Eric Rogers *Physics for the Inquiring Mind* p 196

CHAPTER 27

ERIC ROGERS GETS THE LAST WORD: THE DEMON THEORY OF FRICTION

Eric Rogers

How do you know that it is friction that brings a rolling ball to a stop and not demons? Suppose you answer this, while a neighbour, Faustus, argues for demons. The discussion might run thus:

YOU	I don't believe in demons.
FAUSTUS	I do.
YOU	Anyway, I don't see how demons can make friction.
FAUSTUS	They just stand in front of things and push to stop them from moving.
YOU	I can't see any demons even on the roughest table.
FAUSTUS	They are too small, also transparent.
YOU	But there is more friction on rough surfaces.
FAUSTUS	More demons.
YOU	Oil helps.
FAUSTUS	Oil drowns demons.
YOU	If I polish the table, there is less friction and the ball rolls further.
FAUSTUS	You are wiping the demons off; there are fewer to push.
YOU	A heavier ball experiences more friction.
FAUSTUS	More demons push it; and it crushes their bones more.
YOU	If I put a rough brick on the table I can push against friction with more and more force, up to a limit, and the block stays still, with friction just balancing my push.
FAUSTUS	Of course, the demons push just hard enough to stop you moving the brick; but there is a limit to their strength beyond which they collapse.
YOU	But when I push hard enough and get the brick moving there is friction that drags the brick as it moves along.

237

FAUSTUS	Yes, once they have collapsed the demons are crushed by the brick. It is their crackling bones that oppose the sliding.
YOU	I cannot feel them.
FAUSTUS	Rub your finger along the table.
YOU	Friction follows definite laws. For example, experiment shows that a brick sliding along the table is dragged by friction with a force independent of velocity.
FAUSTUS	Of course, same number of demons to crush, however fast you run over them.
YOU	If I slide a brick along the table again and again, the friction is the same each time. Demons would be crushed in the first trip.
FAUSTUS	Yes, but they multiply incredibly fast.
YOU	There are other laws of friction: for example, the drag is proportional to the pressure holding the surfaces together.
FAUSTUS	The demons live in the pores of the surface: more pressure makes more of them rush out to push and be crushed. Demons act in just the right way to push and drag with the forces you find in your experiments.

By this time Faustus' game is clear. Whatever properties you ascribe to friction he will claim, in some form, for demons. At first his demons appear arbitrary and unreliable; but when you produce regular laws of friction he produces a regular sociology of demons. At that point there is deadlock, with demons and friction serving as alternative names for a set of properties—and each debater is back to his first remark.

You realize that friction has only served you as a name: it has established no link with other properties of matter... And now we can state the full case against demons: they are arbitrary, unreasonable, multitudinous, and over-dressed. We need a special demon with peculiar behaviour to explain each natural event in turn: therefore we need many kinds and vast numbers of them. And we have to clothe them with special behaviours to fit all the facts.

[Abridged from *Physics for the Inquiring Mind* pp 343–5]